The Kids' Guide to Birds of New York

Fun Facts, Activities, and 85 Cool Birds

I0806616

by Stan Tekiela

Adventure Publications
an imprint of AdventureKEEN

DEDICATION

To all the children who enjoy the world of birds as much as I do.

ACKNOWLEDGMENTS

Special thanks to the National Wildlife Refuge System, along with state and local agencies, both public and private, for stewarding the lands that are critical to the many bird species we so love.

Edited by Dan Downing, Sandy Livoti, and Brett Ortler
Cover and book design by Jonathan Norberg
Cover wavy border by ddok/Shutterstock.com
Illustrations by Elleyna Ruud
Range maps produced by Anthony Hertzel

Cover photos by Stan Tekiela. Front: Wild Turkey, Eastern Bluebird, Great Horned Owl, American Goldfinch (main), Wood Duck, and House Finch
Back: Ring-necked Pheasant

All photos by Stan Tekiela except pp. 93 (Eastern Towhee) and 214 (top inset) by **Maslowski Wildlife Productions**; pp. 34 (juvenile) and 174 (in-flight juvenile) by **Brian K. Wheeler**; and p. 96 (top inset) by **Jim Zipp**.

Images used under license from Shutterstock.com:
Agami Photo Agency: 194 (female), 214 (female), 215 (Scarlet Tanager); **cliff collings:** 166 (top inset); **Dennis Jacobsen:** 169 (Eurasian Collared-Dove); **Dr.Pixel:** 228; **Jeff W. Jarrett:** 145 (Ruby-crowned Kinglet); **Jill Schrock:** 198 (winter); **Matthew Jolley:** 208 (female); **Momoe Toyama:** 108 (inset); **Paul Reeves Photography:** 100 (breeding), 158 (breeding); **Protean Visuals:** 204 (inset); **Ray Hennessy:** 100 (top inset); **vagabond54:** 101 (Sanderling); and **Zosma:** 128 (top inset).

To the best of the publisher's knowledge, all photos were of live birds. Some were photographed in a controlled condition.

10 9 8 7 6 5 4 3 2 1

The Kids' Guide to Birds of New York: Fun Facts, Activities, and 85 Cool Birds

Published by Adventure Publications
An imprint of AdventureKEEN
310 Garfield Street South
Cambridge, Minnesota 55008
(800) 678-7006
www.adventurepublications.net

Printed in China
Cataloging-in-Publication data is available from the Library of Congress
ISBN 978-1-64755-507-8 (pbk.); 978-1-64755-508-5 (ebook)

Quick-Flip Color Guide

TABLE OF CONTENTS

Introduction

The Birds

Bird Food Fun for the Family

COOL BIRDS IN NEW YORK

The *Kids' Guide to Birds of New York* is a fun, easy-to-use guide for anyone interested in seeing and identifying birds. As a child, I spent hours of enjoyment watching birds come to a wooden feeder that my father built in our backyard. We were the only family in the neighborhood who fed birds, and we became known as the nature family.

Now, more people feed birds in their backyards than those who go hunting or fishing combined. Not only has it become very popular to feed and watch birds, but young and old alike are also identifying them and learning more about them.

New York is a fantastic state to see all sorts of birds. In fact, more than 450 species are found here. That makes this one of the top places to watch an incredible variety of birds. In this field guide for New York, I'm featuring 85 of the most common of these great birds.

We have marvelous habitats in New York that are perfect for birds. Each of our habitats supports different kinds of birds. There are several mountain ranges that pass through the state. Each of these mountain ranges is covered in trees. This is a great place to see Pine Siskins. In addition to mountains, New York has many significant rivers. In fact, we have more than 70,000 miles of rivers, creeks, and streams. As a result, we have many kinds of water birds such Blue-winged Teal and Wood Ducks.

New York is located in the migratory pathway of many small and large birds. Massive flocks of Canada Geese pass through our state, and birds such as the Palm Warbler arrive here from the warm tropics in the spring.

The weather here also plays a role in the kinds of birds we see. Common Yellowthroats and Baltimore Orioles nest here during summer. Migrating shorebirds, such as Lesser Yellowlegs, pass through New York on their way north. On top of that, backyard birds, most notably Blue Jays and American Goldfinches, enjoy our seasons year-round.

As you can see, New York is a terrific place to watch all kinds of cool birds. It is my sincere hope that you and your family will like watching and feeding birds as much as I did with my family when I was a kid. Let this handy book guide you into a lifetime of appreciating birds and nature.

BODY BASICS OF A BIRD

It's good to know the names of a bird's body parts. The right terminology will help you describe and identify a bird when you talk about it with your friends and family.

The basic parts of a bird are labeled in the illustration below. This drawing is a combination (composite) of several birds and should not be regarded as one particular species.

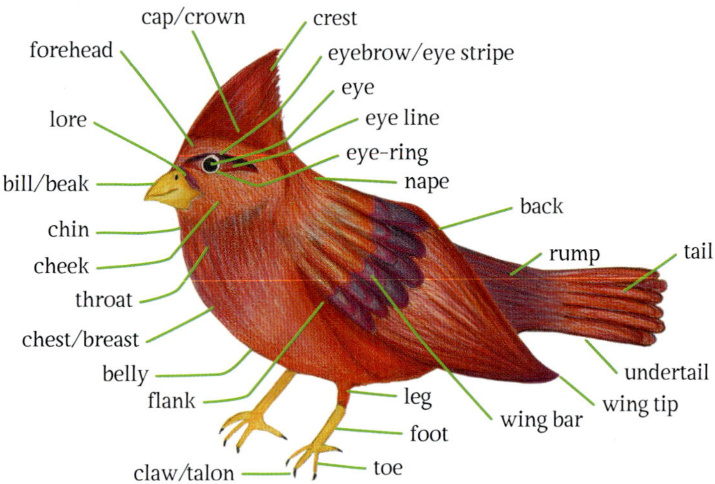

AMAZING NESTS

I am fascinated with bird nests! They are amazing structures that do more than just provide a place for egg laying. Nests create a small climate-controlled environment that's beneficial for both keeping the eggs warm and raising the young after they hatch.

From the high treetops to the ground, there are many kinds of nests. Some are simple, while others are complex. In any case, they function in nearly the same way. Nests help to contain the eggs so they don't roll away. They also help to keep baby birds warm on cold nights, cool on hot days, and dry during rain.

The following illustrations show the major types of nests that birds build in New York.

GROUND **PLATFORM** **CUP** **PENDULOUS** **CAVITY**

A **ground nest** can be a mound of plant materials on the ground or in the water. Some are just a shallow spot scraped in the earth.

A **platform nest** is a cluster of sticks with a depression in the center. It is secured to the platform of a tree fork or to several tree branches.

A **cup nest** has a cupped interior, like a bowl.

A **pendulous nest** is a woven nest that hangs and swings freely, like a pendulum, from a branch.

A **cavity nest** is simply a cavity or hole, usually in a tree.

The first step in nest building is to choose an appropriate site. Each bird species has a unique requirement for this. Some birds, such as Eastern Kingbirds, just need a tree branch. Others, like Eastern Bluebirds, look for a cavity and build the nest inside. Still others, such as Killdeer, search for camouflaged ground to scrape out a nest. Sometimes birds such as Turkey Vultures don't bother building a nest at all if they spot a hard-to-reach cliff or rocky ledge where it will be safe to lay their eggs.

Nest materials usually consist of common natural items found in the area, like sticks or dried grass. Birds use other materials, such as mud or spiderwebs, to glue the materials together.

One of the amazing things about nest construction is that the parents don't need building plans or tool belts. They already know by instinct how to build nests, and they use their beaks and feet as their main tools.

To bring in nesting materials, birds must make many trips back and forth to the nest site. Most use their beaks to hold as much material as possible during each trip. Some of the bigger birds, like Bald Eagles, use larger materials, such as thick sticks and thin branches. They grasp and carry these items with their feet.

Nest building can take two to four days or longer, depending on the species and nest type. The simpler the nest, the faster the construction. Mourning Dove parents, for example, take just a few days to collect one to two dozen sticks for their platform nest. Woodpecker pairs, however, work for upwards of a week to **excavate**, or dig out, a suitable nesting cavity. Large and more complicated platform nests, such as a Bald Eagle nest, may take weeks or even a month to complete, but these can be used for years and are worth the extra effort.

WHO BUILT THAT NEST?

In the majority of bird species, the chief builder is the female. In other species, both the female and the male typically share in the construction equally.

In general, when male and female birds look vastly different, the female does most of the work. When the male and female look alike or appear very similar, they tend to share the tasks of nest building and feeding the young. Alternatively, some species of woodpeckers have a different building plan. When they chisel out a nesting chamber, often the male does more of the work after the female has chosen the site.

ATTRACTING BIRDS WITH FEEDERS

To get more birds to visit your yard, an easy way to invite them is to put out bird feeders. Bird feeders are often as unique as the birds themselves, so the types of feeders you use really depend on the kinds of birds you're trying to attract.

HOPPER TUBE GROUND SUET NECTAR MEALWORM

Hopper feeders are often wooden or plastic. Designed to hold a large amount of seeds, they often have a slender opening along the bottom, which dispenses the seeds. Birds land along the sides and help themselves to the food. Hopper feeders work well as main feeders in conjunction with other types of feeders. They are perfect for offering several kinds of seed mixes for cardinals, finches, nuthatches, chickadees, and more.

Tube feeders with large seed ports and multiple perches are very popular. Often mostly plastic, they tend to be rugged enough to last several years and can be easily cleaned. These feeders are great for black oil sunflower seeds and seed mixes, which are favorites of cardinals and all the other bird species that also visit hopper feeders.

Some tube feeders have small holes, allowing incredibly tiny thistle seeds to be dispensed just a few at a time. Use this kind of feeder to offer Nyjer seed, which will attract various finches.

Other styles of tube feeders have a wire mesh covering with openings large enough for birds to extract one of their favorite foods—peanuts out of the shell. Most birds enjoy peanuts, so these feeders will be some of the most popular in your yard. Another variety of tube feeder has openings large enough for peanuts in the shell. These are also very popular with the birds.

Ground feeders allow a wide variety of birds to access the food. The simplest and easiest feeders to use, they consist of a flat platform with a lip around the edges to keep seeds and corn from spilling out. Some have a roof to keep rain and snow off the food. With or without a roof, drainage holes in the bottom are important. Ground feeders will bring in towhees and many other birds to your backyard, including doves and even mallards if you're near water.

Suet feeders are simply wire cages that hold cakes of **suet**. The wire allows woodpeckers, nuthatches, and other birds to cling securely to the feeder while pecking out chunks of suet. The best suet feeders have a vertical extension at the bottom where a woodpecker can brace its tail and support itself while feeding. These are called tail-prop suet feeders.

Nectar feeders are glass or plastic containers that hold sugar water. These feeders usually have plastic parts that are bright red, a color that is extremely attractive to hummingbirds, but orioles and woodpeckers will also stop for a drink. They often have up to four ports for access to the liquid and yellow bee guards to prevent bees from getting inside.

Mealworm feeders can be very basic—a simple glass or plastic cup or container will do. Pick one with sides tall enough and make sure the material is slippery enough to stop the lively mealworms from crawling out. Bluebirds especially love this wiggly treat!

HOW TO USE THIS GUIDE

Birds move pretty fast, so you don't often get a lot of time to observe them. To help you quickly find the birds in the book, this guide is organized by color. Simply note the most prominent color of the bird you've seen. A male Hairy Woodpecker, for example, is black and white and has a red mark on its head. Since this bird is mostly black and white, you would find it in the black-and-white section.

Within each color section, the birds are organized by size, from small to large. Use the Real Quick sidebar to find the size that your bird appears to be.

When the male and female of a species are different colors (like the Wood Duck pair below), they are shown in their own color sections. In these cases, the opposite sex is included in an inset photo with a page reference so you can easily turn to it.

If you already know the name of the bird you've seen, use the Checklist/Index to get the page number, and flip to it to learn more about the bird.

To further help you with identification, check the range maps to see where and when the bird you have sighted is normally found in New York. Range maps capture our current knowledge of where the birds are during a given year (presence) but do not indicate how many birds are in the area (density). In addition, since birds fly around freely, it's possible to see them outside of their ranges. So please use the maps to get a general idea of where the birds are most likely to be seen.

For more about the information given for each bird in this guide, turn to the Northern Cardinal sample on pgs. 16–17.

While you're learning about birds and identifying them, don't forget to check out the fun-filled things to do starting on pg. 220. Score a big hit with the birds in your yard by creating tasty treats or making your own bird food from the recipes. Put out some nesting materials to help birds build their nests. Consider signing up for a cool citizen science project suitable for the entire family. These are just a few of the activities that are such great fun, you'll want to do them all!

Northern Cardinal

Common name

Look for the black mask

Field markings that help identify the bird

Color tab

Mostly Red

What to look for:
outstanding features; may include other plumages and descriptions

Where you'll find them:
where you're most likely to see the bird

Calls and songs:
songs, calls, and other sounds the bird makes

On the move:
anything about flight, flocks, travel, and other movements

What they eat:
foods the bird eats and the kinds of feeders it visits

Nest:
type of nest; may include nest site, materials, and more

Eggs, chicks, and childcare:
number of eggs, color and marks; **incubation** and feeding duties; may include how many broods

Spends the winter:
where the bird goes when it's cold or when food is scarce

The bold word means it is defined in the glossary

REAL QUICK

Length from head to tail

Size
8–9"

Type of nest the bird calls home

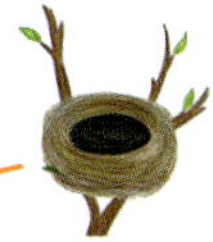

Nest
CUP

Type of feeder the bird generally visits

Feeder
HOPPER

Range map

year-round
summer
migration
winter

After you've seen it, checkmark it

SAW IT!

STAN'S COOL STUFF

Fun and interesting facts about the bird. Information not typically found in other field guides.

Eastern Towhee

Look for the black head

What to look for:
mostly black bird with rusty sides, a white belly, red eyes, and a long black tail with a white tip

Where you'll find them:
shrubby areas with short trees and thick bushes, backyards and parks

Calls and songs:
calls "tow-hee" distinctly; also has a characteristic **call** that sounds like "drink-your-tea"

On the move:
short flights between shrubby areas and heavy **cover**; flashes white wing patches during flight

What they eat:
insects, seeds and fruit; comes to ground feeders

Nest:
cup; Mom constructs the nest

Eggs, chicks, and childcare:
3–4 creamy-white eggs with brown marks; Mom incubates the eggs; Dad and Mom feed the young

Spends the winter:
in New York; moves around in winter

REAL QUICK

Size
7–8"

Nest
CUP

Feeder
GROUND

year-round
summer

SAW IT!

STAN'S COOL STUFF

The towhee is named for its distinctive "tow-hee" call. It hops backward with both feet, raking leaves to find insects and seeds. In southern coastal states, some have white eyes, and others have red eyes.

Brown-headed Cowbird

Look for the brown head

What to look for:
glossy black bird with a chocolate-brown head and a sharp, pointed gray bill

Where you'll find them:
forest edges, open fields, farmlands, and backyards

Calls and songs:
sings a low, gurgling song that sounds like water moving; cowbird young are raised by other bird parents, but they still end up singing and calling like their own parents, whom they've never heard

On the move:
Mom flies quietly to another bird's nest, swiftly lays an egg, then flies quickly away

What they eat:
insects and seeds; visits seed feeders

Nest:
doesn't nest; lays eggs in the nests of other birds

Eggs, chicks, and childcare:
5–7 white eggs with brown marks; the **host** bird incubates any number of cowbird eggs in her nest and feeds the cowbird young along with her own

Spends the winter:
in New York; moves around to find food

REAL QUICK

Size
7½"

Nest
NONE

Feeder
TUBE OR HOPPER

year-round

SAW IT!

STAN'S COOL STUFF

Cowbirds are **brood parasites**, meaning they don't nest or raise their own families. Instead, they lay their eggs in other birds' nests, leaving the host birds to raise their young. Cowbirds have laid their eggs in the nests of more than 200 other bird species.

European Starling

Look for the glittering, iridescent feathers

What to look for:
shiny and **iridescent** purplish-black in spring and summer, speckled in fall and winter; yellow bill in spring, gray in fall; pointed wings and a short tail

Where you'll find them:
lines up with other starlings on power lines; found in all habitats but usually associated with people, farms, suburban yards, and cities

Calls and songs:
mimics the songs of up to 20 bird species; mimics other sounds, even imitating the human voice

On the move:
large family groups gather with blackbirds in fall

What they eat:
bugs, seeds, and fruit; visits seed and **suet** feeders

Nest:
cavity, filled with dried grass; often takes a cavity from other birds

Eggs, chicks, and childcare:
4–6 bluish eggs with brown marks; Mom and Dad sit on the eggs and feed the babies

Spends the winter:
in New York; moves around to find food

REAL QUICK

Size
7½"

Nest
CAVITY

Feeder
TUBE OR HOPPER

year-round

SAW IT!

STAN'S COOL STUFF

The starling is a mimic that can sound like any other bird. It's not a native bird; 100 starlings from Europe were introduced to New York City in 1890–91. Today, European Starlings are one of the most numerous songbirds in the country.

Red-winged Blackbird

Look for the red-and-yellow shoulder patches

What to look for:
black bird with red-and-yellow shoulder patches on upper wings; shoulder patches can be partially or completely covered up

Where you'll find them:
around marshes, wetlands, lakes, and rivers

Calls and songs:
male sings and repeats calls from cattail tops and the surrounding **vegetation**

On the move:
flocks with as many as 10,000 birds gather in autumn, often with other blackbirds

What they eat:
seeds in spring and autumn, insects in summer; visits seed and **suet** feeders

Nest:
cup in a thick stand of cattails over shallow water

Eggs, chicks, and childcare:
3–4 speckled bluish-green eggs; Mom does all the incubating, but both parents feed the babies

Spends the winter:
in New York; moves around in winter to find food

REAL QUICK

Size
8½"

Nest
CUP

Feeder
TUBE OR HOPPER

year-round

SAW IT!

STAN'S COOL STUFF

During autumn and winter, thousands of these birds gather in farm fields, wetlands, and marshes. Come spring, males sing to defend territories and show off their wing patches (**epaulets**) to the females. Later, males can be aggressive when defending their nests.

Common Grackle

Look for the shiny bluish-black head

What to look for:
shiny bluish-black **iridescent** head, a purplish-brown body, and super-bright golden eyes

Where you'll find them:
evergreen trees and shrubs, suburban and urban yards, open fields

Calls and songs:
gives a loud, raspy **call**

On the move:
travels in large flocks with other blackbirds; flight is usually level as opposed to an up-and-down pattern; male holds his tail in a deep V shape

What they eat:
fruit, seeds, and bugs; visits seed and **suet** feeders

Nest:
cup, usually in a **colony** of up to 75 mated pairs

Eggs, chicks, and childcare:
4–5 speckled greenish-white eggs; Mom sits on the eggs; Mom and Dad give food to the babies

Spends the winter:
in southern states; some stay in New York

REAL QUICK

Size
11–13"

Nest
CUP

Feeder
HOPPER

year-round
summer

SAW IT!

STAN'S COOL STUFF

The Common Grackle is a member of the blackbird family. Unlike most birds, it has stronger muscles to open its mouth. The muscles help it to pry apart crevices, where it finds bugs to eat. It's kind of like playing hide-and-seek for its food.

Common Gallinule

Look for the red bill with a yellow tip

What to look for:
nearly black overall with a yellow-tipped red bill, a red forehead, and yellowish-green legs

Where you'll find them:
freshwater marshes and lakes

Calls and songs:
gives a series of fast, high-pitched clucks

On the move:
walks on floating **vegetation** or swims while on the hunt for bugs

What they eat:
insects, snails, seeds, green leaves, fruit, and roots

Nest:
ground nest; Mom and Dad build it with cattails and other wetland plants

Eggs, chicks, and childcare:
2–10 brown eggs with dark marks; Mom and Dad take turns incubating; young usually leave the nest within a few hours after hatching, but they stay with their family for a few months

Spends the winter:
in southern states; some winter on Long Island

REAL QUICK

Size
13–15"

Nest
GROUND

Feeder
NONE

summer
migration
winter

SAW IT!

STAN'S COOL STUFF

This duck-like bird is also called the Pond Chicken. It was once known as the Common Moorhen. Females are known to lay some of their eggs in other Common Gallinule nests. In the water, the young ride around on the backs of the adults.

American Coot

Look for the white bill

What to look for:
gray-to-black with a duck-like white bill, red eyes

Where you'll find them:
in large flocks on open water

Calls and songs:
a unique series of creaks, groans, and clicks

On the move:
bobs head while swimming; takes off from water by scrambling across it with wings flapping; huge flocks of up to 1,000 birds gather for migration; migrates at night

What they eat:
insects and aquatic plants

Nest:
ground nest floating in water, anchored to plants

Eggs, chicks, and childcare:
9–12 speckled pinkish-tan eggs; Mom and Dad sit on the eggs and feed the **hatchlings**

Spends the winter:
in parts of New York, southern states, Mexico, and Central America

REAL QUICK

Size
13–16"

Nest
GROUND

Feeder
NONE

year-round
summer
migration

SAW IT!

STAN'S COOL STUFF

The coot is not a duck. Instead of webbed feet, it has large lobed toes! It's smaller than most other **waterfowl**, and it is a great diver and swimmer. You won't see it flying, but you may spot one trying to escape from a Bald Eagle (pg. 57). It's also called Mud Hen.

American Crow

Look for the glossy black feathers

What to look for:
glossy black all over and a black bill

Where you'll find them:
all habitats—wilderness, rural, suburban, cities

Calls and songs:
a harsh "caw" **call**; imitates other birds and people

On the move:
flaps constantly and glides downward; moves around to find food; gathers in huge communal flocks of more than 10,000 birds during winter

What they eat:
fruit, insects, mammals, fish, and dead carcasses (**carrion**); visits seed and **suet** feeders

Nest:
platform; adds bright or shiny items and often uses the same site every year if a Great Horned Owl (pg. 139) hasn't taken it

Eggs, chicks, and childcare:
4–6 speckled bluish-to-olive eggs; Mom sits on the eggs; Mom and Dad feed the youngsters

Spends the winter:
in New York; moves around in winter to find food, often to city interiors

REAL QUICK

Size
18"

Nest
PLATFORM

Feeder
HOPPER

year-round

SAW IT!

STAN'S COOL STUFF

The crow is one of the smartest of all birds. It's very social and often entertains itself by chasing other birds. It eats roadkill but avoids being hit by vehicles. Some can live as long as 20 years! Crows without mates, called helpers, help to raise the young.

Turkey Vulture

Look for the naked red head

JUVENILE

What to look for:
naked red head and legs and an ivory bill; juvenile has a gray-to-blackish head and bill

Where you'll find them:
in trees, sunning itself with wings outstretched, drying after a rain

Calls and songs:
mostly **mute**, just grunts and groans

On the move:
holds wings in an upright V in flight, teetering from wing tip to wing tip as it soars and hovers

What they eat:
dead carcasses (**carrion**); parents **regurgitate** food for their young

Nest:
no nest, or in a minimal nest on a cliff, in a cave, or even sometimes in a hollow tree trunk

Eggs, chicks, and childcare:
1–3 white eggs with brown marks; Mom and Dad incubate the eggs and feed the baby vultures

Spends the winter:
in southern states, Mexico, Central America, and South America

REAL QUICK

Size
26–32"

Nest
NONE

Feeder
NONE

summer

SAW IT!

STAN'S COOL STUFF

This is one of the few birds with a good sense of smell. It has a strong bill for tearing apart flesh. Unlike hawks and eagles, it has weak feet more suited for walking than grasping wiggly **prey**. The bare head reduces its risk of getting diseases from carcasses.

Double-crested Cormorant

Look for the large, hooked bill

What to look for:
large black waterbird with unusual blue eyes, a long snake-like neck, and large gray bill with a yellow base and hooked tip

Where you'll find them:
usually roosts in large groups in trees near water

Calls and songs:
grunts, pops, and groans—none are pleasant sounds at all!

On the move:
swims underwater to catch fish, holding its wings at its sides; flies in a large V-shaped formation

What they eat:
small fish and aquatic insects

Nest:
platform near or over open water, in a **colony**

Eggs, chicks, and childcare:
3–4 bluish-white eggs; parents take turns sitting on the eggs and feeding the young

Spends the winter:
in southern states, Mexico, and Central America

REAL QUICK

Size
31–35"

Nest
PLATFORM

Feeder
NONE

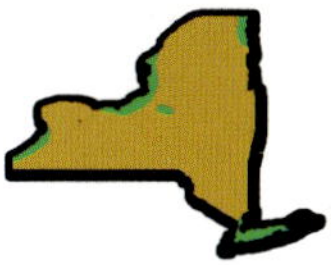

summer
migration

SAW IT!

STAN'S COOL STUFF

This bird's outer feathers are different from its inner ones; the outer feathers soak up water, but its body feathers don't. It opens its wings and uses the sun and wind to dry out. "Double-crested" refers to the two unusual crests on its head, but these aren't often seen.

Downy Woodpecker

Look for the small, short bill

What to look for:
spotted wings, white belly, red mark on the back of the head, and a small, short bill; female lacks a red mark on the head

Where you'll find them:
wherever trees are present

Calls and songs:
repeats a high-pitched "peek-peek" **call**; drums on trees or logs with its bill to announce its territory

On the move:
flies in an up-and-down pattern; makes short flights from tree to tree

What they eat:
insects and seeds; visits **suet** and seed feeders

Nest:
cavity in a dead tree; digs out a perfectly round entrance hole; the bottom of the cavity is wider than the top, and it's lined with fallen woodchips

Eggs, chicks, and childcare:
3–5 white eggs; Mom incubates the eggs; both parents take care of the kiddies

Spends the winter:
in New York

REAL QUICK

Size
6"

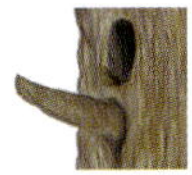

Nest
CAVITY

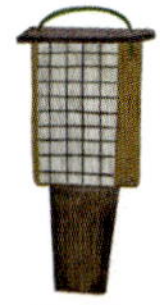

Feeder
SUET

year-round

SAW IT!

STAN'S COOL STUFF

The Downy is abundant and widespread where trees are present. Like other woodpeckers, it pulls insects from tiny places with its long, barbed tongue. It has stiff tail feathers, which help to support it as it clings to trees. During winter, it will roost in a cavity.

Rose-breasted Grosbeak

Look for the rosy-red breast patch

What to look for:
black-and-white bird with a large rosy-red patch in the center of the breast and a large ivory bill

Where you'll find them:
in the woods, usually in pairs around seed feeders

Calls and songs:
male and female sing a rich, robin-like song with a distinctive chip **note** sprinkled in the tune; male is much louder and clearer than the female and sings to her from high up in trees

On the move:
short flights from tree to tree with rapid wing-beats; male flashes white wing patches in flight

What they eat:
insects, seeds, and fruit; visits seed feeders

Nest:
cup in a mature **deciduous** forest, in a small tree

Eggs, chicks, and childcare:
3–5 speckled blue-green eggs; Mom and Dad sit on the eggs and feed the young chicks; juveniles come with the adults to bird feeders

Spends the winter:
in Mexico, Central and South America

REAL QUICK

Size
7–8"

Nest
CUP

Feeder
TUBE OR HOPPER

summer

SAW IT!

STAN'S COOL STUFF

The rose breast patch varies in size and shape among the males. "Grosbeak" refers to the large bill, which the bird uses to crush seeds. In spring, males return before the females. Males become territorial when females arrive and reduce their visits to feeders.

Yellow-bellied Sapsucker

Look for the red chin

What to look for:
checkered back, yellow chest and belly, and a red forehead, crown, and chin; female has a white chin

Where you'll find them:
small woods, forests, suburban, and rural areas, small to medium trees that have rows of sap holes

Calls and songs:
quiet with few vocalizations but will meow like a cat; drums on hollow tree branches irregularly (not in a regular pattern, like other woodpeckers)

On the move:
short up-and-down flights with rapid wingbeats

What they eat:
insects, nutritious sap in trees; visits **suet** feeders

Nest:
cavity that Mom and Dad **excavate**, often in a live tree—but a dead tree will work, too!

Eggs, chicks, and childcare:
5–6 white eggs; Mom and Dad incubate the eggs and feed the **brood**

Spends the winter:
some stay in New York; southern states, Mexico, and Central America

REAL QUICK

Size
8–9"

Nest
CAVITY

Feeder
SUET

year-round
summer
winter

SAW IT!

STAN'S COOL STUFF

Sapsuckers are woodpeckers that drill rows of holes in trees to get to the sap. **Tree sap** is full of minerals and other nutrients. Sapsuckers don't actually suck the sap out of the holes; rather, they lap it with their long, bristly tongues. Oozing sap also attracts bugs to eat.

Hairy Woodpecker

Look for the large bill

What to look for:
spotted wings, white belly, large bill, and red mark on the back of the head; female lacks a red mark

Where you'll find them:
forests and wooded backyards, parks

Calls and songs:
a sharp chirp before landing on feeders; drums on hollow logs, branches, or stovepipes in spring

On the move:
short up-and-down flights from tree to tree with rapid wingbeats

What they eat:
insects, nuts, seeds; visits **suet** and seed feeders

Nest:
cavity; prefers a live tree; excavates a larger, more oval entry than the round hole of the Downy Woodpecker (pg. 39); usually excavates under a branch, which helps to shield the entrance

Eggs, chicks, and childcare:
3–6 white eggs; parents sit on the eggs and bring food to feed their babies

Spends the winter:
in New York

REAL QUICK

Size
9"

Nest
CAVITY

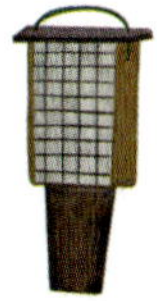

Feeder
SUET

year-round

SAW IT!

STAN'S COOL STUFF

The Hairy is nearly identical to the Downy Woodpecker, but it's larger and has a larger, longer bill. It has a barbed tongue, which it uses to pull out bugs from trees. At the base of its bill, tiny bristle-like feathers protect its nostrils from excavated wood dust.

Red-bellied Woodpecker

Look for the black-and-white striped back

What to look for:
zebra-striped back, white rump, red crown and red nape of neck; female has a light gray crown

Where you'll find them:
shady woodlands, forest edges, and backyards

Calls and songs:
calls a loud "querrr" and a low "chug-chug-chug"

On the move:
rapid wingbeats in flight, going up and down like a roller coaster

What they eat:
spiders and centipedes as well as beetles and other insects; nuts and fruit; visits **suet** and seed feeders

Nest:
cavity in a dead tree; excavates a new hole in last year's tree below the previous cavity

Eggs, chicks, and childcare:
4–5 white eggs; Mom incubates the eggs during the day and Dad takes night duty; both parents feed the baby woodpeckers

Spends the winter:
in New York; moves around to find food

REAL QUICK

Size
9–9½"

Nest
CAVITY

Feeder
SUET

year-round

SAW IT!

STAN'S COOL STUFF

This bird is named for its faint pink belly patch. It excavates dead wood to find bugs to eat and hammers acorns and berries into cracks in trees to store for winter food. The population and range are increasing across the country.

Hooded Merganser

Look for the large white "hood" patch

What to look for:
sleek black-and-white bird with rusty-brown sides; raises its crest "**hood**" to show off a large white patch on each side of its head

Where you'll find them:
wooded areas near ponds, shallow lakes, rivers

Calls and songs:
male gives a deep, rolling **call**; female gives a hoarse quack

On the move:
quick, low flight with fast wingbeats across the water; usually in small groups

What they eat:
small fish, aquatic insects, and **crustaceans** (especially crayfish)

Nest:
cavity; adds a lining of feathers in a natural cavity, an old woodpecker hole or a nest box

Eggs, chicks, and childcare:
10–12 white eggs; Mom incubates the eggs and feeds the babies

Spends the winter:
some stay in New York; southern states

REAL QUICK

Size
16–19"

Nest
CAVITY

Feeder
NONE

year-round
summer

SAW IT!

STAN'S COOL STUFF

The merganser is a small diving bird, not a duck. The male raises his "hood" to show his white head patches. The female will lay eggs in the nests of other mergansers or Wood Ducks (pg. 127). Rarely, she shares a cavity with a Wood Duck, sitting side by side.

Common Goldeneye

Look for the large white spot on the face

pg. 129

What to look for:
white body duck with puffy round green-black head and large white patch at base of black bill, yellow eye

Where you'll find them:
on small and large bodies of water, fresh and saltwater

Calls and songs:
male gives a high-pitched "yeee-pe;" wings whistle boldly when taking flight

On the move:
very fast straight flight, constantly flapping

What they eat:
aquatic plants, insects, fish, and mollusks

Nest:
cavity; adds a lining of feathers in an old woodpecker cavity or natural cavity in a tree

Eggs, chicks, and childcare:
8–10 light green eggs; Mom incubates the eggs and leads the babies to food

Spends the winter:
some stay in New York; southern states and Mexico; moves around to find open water

REAL QUICK

Size
18–20"

Nest
CAVITY

Feeder
NONE

summer
migration
winter

SAW IT!

STAN'S COOL STUFF

A small duck of both fresh and saltwater. One of the last duck species to migrate each fall. A diving duck that feeds on aquatic insects. Fast flying bird reaching around 45 mph.

Pileated Woodpecker

Look for the bright red crest

What to look for:

bright red crest that looks like a hat; bright red forehead and mustache, and a black back; female has a black forehead and lacks a red mustache

Where you'll find them:

prefers areas with lots of woodland

Calls and songs:

drums on hollow branches, chimneys, and such to announce territory; loud, rapid "cuk-cuk-cuk" calls carry over a long distance

On the move:

white leading edge of wings flashes brightly during flight

What they eat:

insects (especially its favorite, carpenter ants); visits **suet** feeders and feeders with peanuts

Nest:

cavity in a dead or live tree trunk

Eggs, chicks, and childcare:

3–5 white eggs; Mom and Dad sit on the eggs and **regurgitate** bugs to feed the youngsters

Spends the winter:

doesn't **migrate**; stays in New York year-round

REAL QUICK

Size
19"

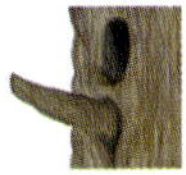

Nest
CAVITY

Feeder
SUET

year-round

SAW IT!

STAN'S COOL STUFF

This is our largest woodpecker. It's shy, despite its size. It digs oval holes up to a few feet long in tree trunks, looking for bugs to eat. You'll see large wood chips at the base of those trees. The young come out of the nest looking and sounding just like the adults.

Osprey

Look for the dark line through the eyes

What to look for:
white chest, belly, and head, with a dark eye line

Where you'll find them:
always near water, from rivers to wetlands

Calls and songs:
a high-pitched, whistle-like **call**, often given in flight as a warning

On the move:
can hover for a few seconds before diving to catch a fish; carries fish in a headfirst position in flight for better aerodynamics

What they eat:
fish

Nest:
platform made with twigs; on a raised wooden platform, man-made tower, or in a tall dead tree

Eggs, chicks, and childcare:
2–4 white eggs with brown marks; parents sit on the eggs and feed the **hatchlings**

Spends the winter:
in southern states, Mexico, Central and South America

REAL QUICK

Size
21–24"

Nest
PLATFORM

Feeder
NONE

summer
migration

SAW IT!

STAN'S COOL STUFF

The Osprey is the only species in its family. It is the only **raptor** that plunges feetfirst into the water to catch fish. Bald Eagles (pg. 57) will harass it for its catch. At one time, it was almost extinct. It was reintroduced to many regions, and populations are now stable.

Bald Eagle

Look for the white head

What to look for:
white head and tail, curved yellow bill and yellow feet; juvenile has white speckles and a gray bill

Where you'll find them:
often near water; likes open areas with daily food

Calls and songs:
weak, high-pitched screams, one after another

On the move:
a spectacular aerial mating **display**: one eagle flips upside down and locks talons with another; both fall, tumbling to earth, then break apart and fly off

What they eat:
fish, **carrion** (dead rabbits and squirrels), birds (mainly ducks)

Nest:
massive platform of sticks, usually in a tree; nests used for many years can weigh up to 1,000 pounds

Eggs, chicks, and childcare:
2–3 off-white eggs; Mom and Dad share all duties

Spends the winter:
in New York; moves around in winter to find open water and fish

REAL QUICK

Size
31–37"

Nest
PLATFORM

Feeder
NONE

year-round
winter

SAW IT!

STAN'S COOL STUFF

Bald Eagles nearly became extinct, but they're doing well now. Their wingspan is huge, stretching out to 7½ feet! They return to the same nest and add more sticks each year, enlarging it over time. The heads and tails of juveniles turn white at 4–5 years.

Indigo Bunting

Look for the vibrant blue feathers

What to look for:
vibrant blue with scattered dark marks on the wings and tail; **plumage** gleams in direct sunlight and appears dull on cloudy days or in shade

Where you'll find them:
woodland edges, where it feasts on insects; parks and yards

Calls and songs:
male often sings from treetops to attract a mate; female is quiet

On the move:
migrates at night in flocks of 5–10 birds

What they eat:
insects, seeds, and fruit; only visits seed feeders early in spring, when bugs are in short supply

Nest:
cup in a small tree or shrub, low to the ground

Eggs, chicks, and childcare:
3–4 pale blue eggs; Mom sits on the eggs and attends to the young

Spends the winter:
in southern Florida, Mexico, Central and South America

REAL QUICK

Size
5½"

Nest
CUP

Feeder
HOPPER

summer

SAW IT!

STAN'S COOL STUFF

This male is actually gray! Like Blue Jays (pg. 67) and other blue birds, there's no blue pigment in the feathers. Sunlight **refraction** in the structure of the feathers makes them look blue. Males **molt** in autumn and look like the brown females during winter.

Tree Swallow

Look for the white chin and chest

What to look for:
blue-green bird with a white chin, chest, and belly, and long, pointed wings

Where you'll find them:
ponds, lakes, rivers, and farm fields

Calls and songs:
gives a series of gurgles and chirps; chatters when upset or threatened

On the move:
flies back and forth across fields, feeding on bugs; uses rapid wingbeats, and then glides; family units gather in large flocks for migration

What they eat:
insects—big and small

Nest:
cavity; adds grass and lines it with feathers; uses an old woodpecker hole or a wooden nest box

Eggs, chicks, and childcare:
4–6 white eggs; Mom sits on the eggs; Mom and Dad bring bugs to feed the babies

Spends the winter:
in Mexico and Central America

REAL QUICK

Size
5–6"

Nest
CAVITY

Feeder
NONE

summer

SAW IT!

STAN'S COOL STUFF

This swallow is a good bird to have around because it eats many bugs. You can attract it with a nest box, but it will compete with Eastern Bluebirds (pg. 65) for the cavity. It finds dropped feathers to line its nest and plays with them on its way back to the nest.

Barn Swallow

Look for the deeply forked tail

What to look for:
sleek blue-black back, rusty chin, cinnamon belly, and a long, deeply forked tail

Where you'll find them:
wetlands, farms, suburban yards, and parks

Calls and songs:
gives a twittering **warble** that's followed by a rapid mechanical sound

On the move:
flaps continuously, often low over land or water; unlike other swallows, it rarely glides

What they eat:
bugs—especially beetles, wasps (caught carefully), and flies

Nest:
cup made of mud; brings in up to 1,000 beak-loads of mud to build nest, often on a building; usually it's in a **colony** of 4–6 birds; sometimes nests alone

Eggs, chicks, and childcare:
4–5 white eggs with brown marks; Mom sits on the eggs, and Mom and Dad feed the chicks

Spends the winter:
in South America

REAL QUICK

Size
7"

Nest
CUP

Feeder
NONE

summer

SAW IT!

STAN'S COOL STUFF

The Barn Swallow is the only swallow in New York with a deeply forked tail. It drinks while flying low over water, and it sips the waterdrops on wet leaves. It bathes when it flies through rain or sprinklers. Usually, it nests on a barn, house, or under a bridge.

Eastern Bluebird

Look for the rusty-red chest

MALE

FEMALE

What to look for:
sky-blue head, back, wings, and tail, with a rusty-red chest and white belly; female is grayer

Where you'll find them:
open habitats (prefers farm fields, pastures and roadsides), forest edges, parks, and yards

Calls and songs:
male repeats a distinctive "chur-lee chur chur-lee" song mostly in spring as he displays to the female

On the move:
short flights from tree to tree; often perches in trees or on posts, dropping to ground to grab bugs

What they eat:
insects, fruit; visits mealworm and **suet** feeders

Nest:
cavity; adds a soft lining in an old woodpecker hole or a bluebird nest box

Eggs, chicks, and childcare:
4–5 pale blue eggs; Mom incubates the eggs, and Dad and Mom feed the kids; 2 broods per year

Spends the winter:
in New York; moves just far enough south to avoid the harshest parts of winter

REAL QUICK

Size
7"

Nest
CAVITY

Feeder
MEALWORM

year-round
summer
migration

SAW IT!

STAN'S COOL STUFF

The bluebird is a cousin of the American Robin (pg. 165). It was nearly eliminated due to a lack of tree cavities, but it's thriving now thanks to people who have put up bluebird nest boxes. The young of the first **brood** sometimes help care for the second brood.

Blue Jay

Look for the large crest

What to look for:
vivid blue bird with a black **necklace**; a large crest, which the jay raises and lowers at will

Where you'll find them:
in the woods and all around your backyard

Calls and songs:
loud, noisy, and mimics other birds; screams like a hawk around feeders to scare away other birds

On the move:
carries seeds and nuts in a pouch under its tongue during flight

What they eat:
insects, fruit, seeds, nuts, bird eggs, and babies in other nests; visits seed feeders, ground feeders with corn, and any feeder with peanuts

Nest:
cup of twigs in a tree, near the main trunk

Eggs, chicks, and childcare:
4–5 speckled green-to-blue eggs; Mom sits on the eggs; Mom and Dad feed the little ones

Spends the winter:
in New York; moves around to find an abundant source of food

REAL QUICK

Size
12"

Nest
CUP

Feeder
HOPPER

year-round

SAW IT!

STAN'S COOL STUFF

Blue Jays are very intelligent. They store food in hiding places, called caches, to eat later. They can remember where they hid thousands of nuts! Jays will imitate hawks to scare off other birds at feeders before they land to get the food.

Belted Kingfisher

Look for the large, ragged crest

What to look for:
broad blue band on a white chest, ragged crest; female has a rusty band below her blue band

Where you'll find them:
rarely away from water; usually at banks of rivers, lakes, and large streams

Calls and songs:
gives a loud **call** that sounds like a machine gun rattling; mates know each other's call

On the move:
flashes white wing patches during flight

What they eat:
small fish

Nest:
cavity in the bank of a river, lake, or cliff; digs a tunnel up to 4 feet long to the nest chamber

Eggs, chicks, and childcare:
6–7 white eggs; Mom and Dad sit on the eggs and feed fish to their youngsters

Spends the winter:
in southern states, Mexico, Central and South America

REAL QUICK

Size
12–14"

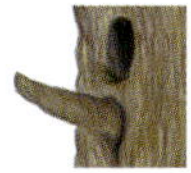

Nest
CAVITY

Feeder
NONE

year-round
summer

SAW IT!

STAN'S COOL STUFF

Belted Kingfishers perch near water and dive in headfirst to catch fish. Parents drop dead fish into the water to teach their young to dive. Kingfishers have short legs with two toes fused together. This helps a lot when they dig (**excavate**) a burrow for nesting.

Chipping Sparrow

Look for the rusty crown

What to look for:
gray-brown bird with a clear gray chest, white eyebrows, thin black eye line, and a rusty crown

Where you'll find them:
forest edges, suburban yards, parks, openings in the forest

Calls and songs:
male gives a very long, dry, and fast "chip" **call**, repeating it over and over

On the move:
gathers and feeds in large family groups in preparation for migration; flocks of 20–30 birds **migrate** at night

What they eat:
insects and seeds; comes to ground feeders

Nest:
cup, usually lining it with animal hair; builds nest low in dense shrubs

Eggs, chicks, and childcare:
3–5 speckled blue-green eggs; Mom incubates the eggs; Mom and Dad share the childcare

Spends the winter:
in southern states, Mexico, and Central America

REAL QUICK

Size
5"

Nest
CUP

Feeder
GROUND

summer

SAW IT!

STAN'S COOL STUFF

This sparrow is a common garden or backyard bird, often dining on dropped seeds beneath feeders. Usually, it isn't very afraid of people, so you may be able to come near it before it flies away. It is named for its quick "chip" call, and it's also known just as Chippy.

Common Redpoll

Look for the bright red crown

What to look for:
heavily streaked back, a splash of raspberry-red on the chest, a red crown, and a black spot on the chin; female lacks red on the chest

Where you'll find them:
coniferous to **deciduous** forests, open fields

Calls and songs:
electric-sounding zipping **call**, given in long strings lasting upwards of 30 seconds; also gives a nasal, rising whistle

On the move:
small to large flocks fly from location to location, wheeling around in curves and circles in the sky before landing at feeders

What they eat:
seeds and insects; comes to seed feeders in winter

Nest:
cup; Mom builds it

Eggs, chicks, and childcare:
4–5 pale green eggs with purple marks; Mom incubates the eggs; Mom and Dad feed the chicks

Spends the winter:
in New York, flying in from Canada

REAL QUICK

Size
5"

Nest
CUP

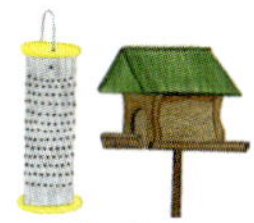

Feeder
TUBE OR HOPPER

winter

SAW IT!

STAN'S COOL STUFF

Redpolls are winter visitors in New York. They move around in flocks of up to 100 birds, but in some years, they are not seen at all. While they're here, they bathe in snow or open water. Like Black-capped Chickadees (pg. 147), they can be tamed and hand-fed.

Pine Siskin

Look for the yellow on the wings

What to look for:
brown with a heavily streaked back, chest, and belly, yellow wing bars and yellow at base of tail

Where you'll find them:
coniferous to **deciduous** forests, open fields

Calls and songs:
gives a series of high-pitched, wheezy calls; also gives a wheezing **twitter**

On the move:
moves around to visit feeders in flocks of up to 20 or more birds, often with other finch species; flashes yellow wing markings in flight

What they eat:
seeds, bugs; visits seed (especially thistle) feeders

Nest:
cup; builds nest in a conifer near the end of a branch, where needles are dense

Eggs, chicks, and childcare:
3–4 speckled greenish-blue eggs; Mom incubates the eggs; Mom and Dad bring food to the kiddies

Spends the winter:
in New York; moves around in search of food during winter

REAL QUICK

Size
5"

Nest
CUP

Feeder
TUBE OR HOPPER

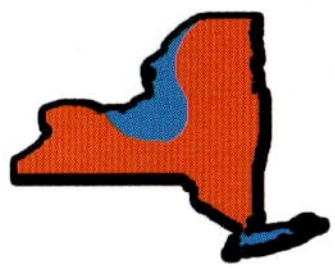

year-round
winter

SAW IT!

STAN'S COOL STUFF

The Pine Siskin is a finch that breeds in small groups. Nests in the group are often only a few feet apart. The male feeds the female during **incubation.** Juveniles have a yellow tint on their chests and chins, but they lose this by late summer of their first year.

House Finch

Look for the heavily streaked chest

FEMALE

MALE
pg. 193

What to look for:
brown bird with heavy streaks on a white chest

Where you'll find them:
forests, city, and suburban areas, around homes, parks, and farms

Calls and songs:
male sings a loud, cheerful warbling song

On the move:
moves around in small family units; never travels long distances

What they eat:
seeds, fruit, and leaf buds; comes to seed feeders and feeders with a glop of grape jelly

Nest:
cup, but occasionally in a cavity; likes to nest in a hanging flower basket or on a front door wreath

Eggs, chicks, and childcare:
4–5 pale blue eggs, lightly marked; Mom sits on the eggs and Dad feeds her while she incubates; Mom and Dad feed the **brood**

Spends the winter:
in New York; moves around to find food

REAL QUICK

Size
5"

Nest
CUP

Feeder
TUBE OR HOPPER

year-round

SAW IT!

STAN'S COOL STUFF

The House Finch is very social and can be the most common bird at feeders. It was introduced to New York from the western U.S. in the 1940s. Now, it's found all across the country. Unfortunately, it suffers from an eye disease that causes the eyes to crust over.

House Wren

Look for the slightly curved bill

What to look for:
brown bird with light brown marks on the wings and tail, a slightly curved brown bill

Where you'll find them:
brushy yards, woodlands, forest edges, and parks

Calls and songs:
sings a lot; during the mating season, it sings from dawn to dusk

On the move:
short flights from protective bushes; holds its tail up briefly after landing

What they eat:
insects, spiders, and snails

Nest:
cavity in a tree or birdhouse; easily attracted to a nest box; builds a twiggy nest in spring and lines it with pine needles and grass

Eggs, chicks, and childcare:
4–6 tan eggs with brown marks; Mom and Dad incubate the eggs and raise the chicks

Spends the winter:
in southern states and Mexico

REAL QUICK

Size
5"

Nest
CAVITY

Feeder
NONE

summer

SAW IT!

STAN'S COOL STUFF

The male chooses several nest cavities and puts a few small twigs in each. The female selects one cavity and then fills it with short twigs. Often, she will have trouble fitting long twigs through the entrance hole, but she'll try again and again until she's successful.

Carolina Wren

Look for the bold white eyebrows

What to look for:
orange-yellow chest and belly, a white throat, bold white eyebrows, and a stubby tail, often held up

Where you'll find them:
brushy yards and woodlands

Calls and songs:
sings year-round; male sings up to 40 song types, singing one song repeatedly before switching to another; female also sings, resulting in duets

On the move:
short, fast flights, often perching high up to sing before flying again

What they eat:
insects, fruit, and few seeds; comes to **suet** and mealworm feeders

Nest:
cavity; nests in birdhouses and in unusual places, like mailboxes, car bumpers, and broken taillights

Eggs, chicks, and childcare:
4–6 white (sometimes pink) eggs with brown marks; Mom incubates; parents feed the babies

Spends the winter:
doesn't **migrate**; stays in New York year-round

REAL QUICK

Size
$5\frac{1}{2}$"

Nest
CAVITY

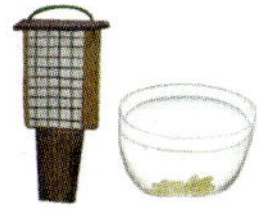

Feeder
SUET OR MEALWORM

year-round

SAW IT!

STAN'S COOL STUFF

Carolina Wrens have a long-term **pair bond**. Mated pairs stay with each other in their territory all year long. They can have up to three broods per year. The male often takes over feeding the young when the female nests again.

Indigo Bunting

Look for the faint blue on the wings

What to look for:
light brown bird with faint wing bars and faint blue on the wings

Where you'll find them:
woodland edges, where it feasts on insects; parks and yards

Calls and songs:
female is quiet; male often sings from treetops to attract a mate

On the move:
migrates at night in flocks of 5–10 birds

What they eat:
insects, seeds, and fruit; only visits seed feeders early in spring, when bugs are in short supply

Nest:
cup in a small tree or shrub, low to the ground

Eggs, chicks, and childcare:
3–4 pale blue eggs; Mom sits on the eggs and attends to the young

Spends the winter:
in southern Florida, Mexico, Central and South America

REAL QUICK

Size
5½"

Nest
CUP

Feeder
HOPPER

summer

SAW IT!

STAN'S COOL STUFF

The female Indigo Bunting is secretive and plain, so usually only males are noticed. Females and juveniles return in spring after the males arrive, typically to their nest sites from the previous year. Juveniles move to areas within a mile of their birthplaces.

Dark-eyed Junco

Look for the ivory-to-pink bill

pg. 153

Mostly Brown

What to look for:
plump bird with a tan-to-brown chest, head and back, a white belly, and a tiny ivory-to-pink bill

Where you'll find them:
on the ground in small flocks with other juncos and sparrows

Calls and songs:
a beautiful, loud musical **trill** lasting 2–3 seconds

On the move:
outermost tail feathers are white and appear as a white V during flight

What they eat:
seeds (scoffs down many weed seeds) and insects; visits ground and seed feeders

Nest:
cup on the ground in a wide variety of habitats; female chooses a well-hidden nest site

Eggs, chicks, and childcare:
3–5 white eggs with reddish-brown marks; Mom incubates the eggs; Dad and Mom feed the babies

Spends the winter:
stays in parts of New York; moves around in winter

REAL QUICK

Size
5½"

Nest
CUP

Feeder
GROUND

year-round
winter

SAW IT!

STAN'S COOL STUFF

The junco is one of our most common winter birds. Females **migrate** farther south than males. This round, dark-eyed bird uses both feet at the same time to "**double-scratch**" the ground, exposing seeds and insects to eat.

Song Sparrow

Look for the central dark spot on the chest

What to look for:
brown bird with heavy dark streaking on the chest, some connecting to form a central dark spot

Where you'll find them:
scrubby areas, backyards, open fields, forest edges

Calls and songs:
sings constantly, repeating a loud, clear song every couple of minutes; starts a song with 3 separate notes, then finishes up with a **trill**; sings from thick shrubs and defends its territory with songs

On the move:
rarely in **flocks** (unlike many sparrow species)

What they eat:
insects and seeds; rarely visits ground feeders with seeds

Nest:
cup; Mom is the sole designer and builder

Eggs, chicks, and childcare:
3–4 pale blue-to-green eggs with reddish-brown splotches; Mom sits on the eggs, and Mom and Dad give food to the **hatchlings**; 2 **broods** per year

Spends the winter:
in New York; moves around in search of food

REAL QUICK

Size
5–6"

Nest
CUP

Feeder
GROUND

year-round
summer

SAW IT!

STAN'S COOL STUFF

This bird has many varieties, but they all have a central dark spot on their chests, and they sing unique songs. Song Sparrows are ground feeders. They make little scraping hops backward with both feet to **double-scratch** and uncover seeds to eat.

House Sparrow

Look for the black throat patch

MALE

FEMALE

What to look for:
brown back, gray belly and crown, large black patch from throat to chest; female is light brown with distinct light eyebrows, lacks a throat patch

Where you'll find them:
just about any **habitat**, from cities to farms

Calls and songs:
one of the first birds heard in cities during spring

On the move:
nearly always in small flocks

What they eat:
seeds, insects, and fruit; comes to seed feeders

Nest:
cavity; uses dried grass, scraps of plastic, paper and whatever else is available to construct an oversized domed cup within the cavity

Eggs, chicks, and childcare:
4–6 white eggs with brown marks; Mom sits on the eggs; Mom and Dad feed the little ones

Spends the winter:
in New York; moves around to find food

REAL QUICK

Size
6"

Nest
CAVITY

Feeder
TUBE OR HOPPER

year-round

SAW IT!

STAN'S COOL STUFF

The House Sparrow is very comfortable being around people. It was introduced to Central Park in New York City from Europe in 1850. It adjusted to nearly all habitats and now is seen across North America. Populations are decreasing in the U.S. and worldwide.

White-throated Sparrow

Look for the light stripes on the head

WHITE-STRIPED

TAN-STRIPED

What to look for:
striped head, white or tan throat patch, and small yellow **lores** between the eyes

Where you'll find them:
bogs, evergreen, and leafy forests, under feeders

Calls and songs:
sings a wonderful song all year and can even be heard at night, sounding like "oh-Canada, Canada"

On the move:
often hangs around on the ground with other sparrows during winter

What they eat:
insects, seeds, and fruit; comes to ground feeders

Nest:
cup on the ground under a small tree

Eggs, chicks, and childcare:
4–6 greenish, bluish, or creamy-white eggs with reddish-brown marks; Mom incubates the eggs; Mom and Dad both take care of the babies

Spends the winter:
in New York; southern states

REAL QUICK

Size
6-7"

Nest
CUP

Feeder
GROUND

year-round
summer
winter

SAW IT!

STAN'S COOL STUFF

This bird has two color variations: white-striped and tan-striped. Both variations mate with each other. Both variations also sing, except for the tan-striped females. This is odd, and scientists aren't sure why those females don't sing. Maybe you can figure it out.

Eastern Towhee

Look for the rusty sides

FEMALE

MALE
pg. 19

What to look for:
light brown bird with rusty sides, a white belly, red eyes, and a long brown tail with a white tip

Where you'll find them:
shrubby areas with short trees and thick bushes, backyards and parks

Calls and songs:
calls "tow-hee" distinctly; also has a characteristic **call** that sounds like "drink-your-tea"

On the move:
short flights between shrubby areas and heavy **cover**; flashes white wing patches during flight

What they eat:
insects, seeds, and fruit; comes to ground feeders

Nest:
cup; Mom constructs the nest

Eggs, chicks, and childcare:
3–4 creamy-white eggs with brown marks; Mom incubates the eggs; Dad and Mom feed the young

Spends the winter:
in New York, southern states, and Mexico

REAL QUICK

Size
7–8"

Nest
CUP

Feeder
GROUND

year-round
summer

SAW IT!

STAN'S COOL STUFF

The towhee is named for its distinctive "tow-hee" call. It hops backward with both feet, raking leaves to find insects and seeds. In southern coastal states, some have white eyes and others have red eyes. Usually heard well before it is seen.

Rose-breasted Grosbeak

Look for the large white eyebrows

What to look for:
heavily streaked bird with orange-to-yellow wing linings, bold white eyebrows, and a large ivory bill

Where you'll find them:
in the woods, usually in pairs around seed feeders

Calls and songs:
female and male sing a rich, robin-like song with a distinctive chip **note** sprinkled in the tune; male is much louder and clearer than the female and sings to her from high up in trees

On the move:
short flights from tree to tree with rapid wing-beats; male flashes white wing patches in flight

What they eat:
insects, seeds, and fruit; visits seed feeders

Nest:
cup in a mature **deciduous** forest, in a small tree

Eggs, chicks, and childcare:
3–5 speckled blue-green eggs; Mom and Dad sit on the eggs and feed the young chicks; juveniles come with the adults to bird feeders

Spends the winter:
in Mexico, Central and South America

REAL QUICK

Size
7-8"

Nest
CUP

Feeder
TUBE OR HOPPER

summer

SAW IT!

STAN'S COOL STUFF

The common name "Grosbeak" refers to the large bill, which the bird uses to crush seeds. In spring, females return to New York several days after the males arrive. After the young leave the nest (**fledge**), they come to seed feeders with the adults.

Brown-headed Cowbird

Look for the pointed gray bill

What to look for:
brown bird with a sharp, pointed gray bill

Where you'll find them:
forest edges, open fields, farmlands, and backyards

Calls and songs:
sings a low, gurgling song that sounds like water moving; cowbird young are raised by other bird parents, but they still end up singing and calling like their own parents, whom they've never heard

On the move:
Mom flies quietly to another bird's nest, swiftly lays an egg, then flies quickly away

What they eat:
insects and seeds; visits seed feeders

Nest:
doesn't nest; lays eggs in the nests of other birds

Eggs, chicks, and childcare:
5–7 white eggs with brown marks; the **host** bird incubates any number of cowbird eggs in her nest and feeds the cowbird young along with her own

Spends the winter:
in New York

REAL QUICK

Size
7½"

Nest
NONE

Feeder
TUBE OR HOPPER

year-round

SAW IT!

STAN'S COOL STUFF

Cowbirds are **brood parasites**, meaning they don't nest or raise their own families. Instead, they lay their eggs in other birds' nests, leaving the host birds to raise their young. Cowbirds have laid their eggs in the nests of more than 200 other bird species.

Cedar Waxwing

Look for the waxy-looking red wing tips

What to look for:
sleek bird with a pointed crest, black mask, light yellow belly, and waxy-looking red wing tips; tail has a bold yellow tip; juvenile lacks red wing tips

Where you'll find them:
treetops, forest edges, in trees with fruit

Calls and songs:
constantly makes a high-pitched "sreee" whistling sound while it's perched or in flight

On the move:
flies in flocks of 5–100 birds; moves from area to area, looking for berries

What they eat:
berry-like cedar cones, fruit, seeds, and insects

Nest:
cup; Mom and Dad construct it together

Eggs, chicks, and childcare:
4–6 pale blue eggs with brown marks; Mom sits on the eggs; Mom and Dad feed the little ones

Spends the winter:
in New York; wanders around in search of available food supplies

REAL QUICK

Size
$7\frac{1}{2}$"

Nest
CUP

Feeder
NONE

year-round

SAW IT!

STAN'S COOL STUFF

The waxwing is named for its waxy-looking red wing tips and for the cedar's small, blueberry-like cones that it likes to eat. Before berries are abundant, it eats bugs. The young obtain the mask after their first year of life and red wing tips after their second year.

Sanderling

Look for the short black bill

What to look for:
rusty head, chest, and back with a white belly, and black legs and bill; white on the wings, seen only in flight

Where you'll find them:
on sandy beaches

Calls and songs:
male gives a high-pitched **call** in flight when he displays to the female

On the move:
when waves retreat at the beach, groups run out to feed; often hops away from people on one leg; performs a distraction **display** when threatened

What they eat:
insects, crabs, worms and small **mollusks**

Nest:
ground nest; Dad builds it

Eggs, chicks, and childcare:
3–4 greenish-olive eggs with brown marks; the parents do the **incubation** and feed the kids

Spends the winter:
on East and Gulf Coasts, in the Caribbean, Mexico, Central and South America

REAL QUICK

Size
8"

Nest
GROUND

Feeder
NONE

migration
winter

SAW IT!

STAN'S COOL STUFF

The Sanderling is one of the most common shorebirds in New York. It has breeding **plumage** from April to August and winter plumage from August to April. To rest, it stands on one leg and tucks the other leg into its belly feathers. It nests in the Arctic.

Northern Cardinal

Look for the reddish bill

What to look for:
tan-to-brown bird with a black mask and a large reddish bill; juvenile has a blackish-gray bill

Where you'll find them:
wide variety of habitats, including backyards and parks; usually likes thick **vegetation**

Calls and songs:
calls "whata-cheer-cheer-cheer" in spring; both female and male sing and give chip notes all year

On the move:
short flights from **cover** to cover, often landing on the ground

What they eat:
loves sunflower seeds and enjoys insects, fruit, peanuts, and **suet**; visits seed feeders

Nest:
cup of twigs and bark strips, often low in a tree

Eggs, chicks, and childcare:
3–4 speckled bluish-white eggs; Mom and Dad share the incubating and feeding duties

Spends the winter:
doesn't **migrate**; gathers with other cardinals and moves around to find good sources of food

REAL QUICK

Size
8-9"

Nest
CUP

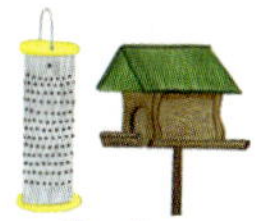

Feeder
TUBE OR HOPPER

year-round

SAW IT!

STAN'S COOL STUFF

Cardinals are sunbathers! Sometimes they stretch out in the sun, spreading their wings and fanning their tails. They are the first to arrive at feeders in the morning and the last to leave before dark. Females sing as loud as the males—only a few bird species do this.

Red-winged Blackbird

Look for the white eyebrows

What to look for:
heavily streaked with a pointed brown bill and white (sometimes yellow) eyebrows

Where you'll find them:
around marshes, wetlands, lakes, and rivers

Calls and songs:
male sings and repeats calls from cattail tops and the surrounding **vegetation**

On the move:
flocks with as many as 10,000 birds gather in autumn, often with other blackbirds

What they eat:
seeds in spring and autumn, insects in summer; visits seed and **suet** feeders

Nest:
cup in a thick stand of cattails over shallow water

Eggs, chicks, and childcare:
3–4 speckled bluish-green eggs; Mom does all the incubating, but both parents feed the babies

Spends the winter:
in New York; moves around in winter to find food

REAL QUICK

Size
8½"

Nest
CUP

Feeder
TUBE OR HOPPER

year-round

SAW IT!

STAN'S COOL STUFF

During autumn and winter, thousands of these birds gather in farm fields, wetlands, and marshes. Come spring, males sing to defend territories and show off their wing patches (**epaulets**) to the females. Later, males can be aggressive when defending their nests.

American Kestrel

Look for the black lines on the face

What to look for:
rusty back, blue-gray wings, spotted chest, two black lines on the face, a wide black band on the tip of tail; female has rusty wings, dark tail bands

Where you'll find them:
open fields, prairies, farm fields, along highways

Calls and songs:
loud series of high-pitched "klee-klee-klee" calls

On the move:
hovers in midair near roads, then dives for **prey**; pumps tail up and down after landing on a perch

What they eat:
bugs (especially grasshoppers), small animals and birds, reptiles

Nest:
cavity in a tree or wooden nest box; doesn't add nesting material

Eggs, chicks, and childcare:
4–5 white eggs with brown marks; parents take turns sitting on the eggs and feeding the babies

Spends the winter:
in New York

REAL QUICK

Size
9–11"

Nest
CAVITY

Feeder
NONE

SAW IT!

STAN'S COOL STUFF

The kestrel is a small falcon that perches nearly upright. The male and female have different markings—this is unusual for a **raptor**. It can see **ultraviolet light**. That ability helps it find mice and other prey by their urine, which glows bright yellow in ultraviolet light.

Lesser Yellowlegs

Look for the long yellow legs

What to look for:
typical sandpiper-type bird with a brown back and wings, streaked white chest; thin, straight black bill; long yellow legs

Where you'll find them:
wetlands, ponds, along rivers and lakes, flooded fields

Calls and songs:
many short high whistles in pairs of two

On the move:
stalks the edge of water

What they eat:
aquatic insects, tiny fish

Nest:
ground; Mom builds it

Eggs, chicks, and childcare:
3–4 yellowish eggs with brown marks; Mom and Dad sit on the eggs; Mom and Dad lead the young to food

Spends the winter:
in southern coastal states, Mexico, Central America, and South America

REAL QUICK

Size
10–12"

Nest
GROUND

Feeder
NONE

migration

SAW IT!

STAN'S COOL STUFF

It nests in Canada but passes through New York twice a year. Often seen in flooded fields or shallow marshes. Was once nearly killed off due to uncontrolled hunting.

Killdeer

Look for the two black neckbands

What to look for:
brown back, white belly, two black bands around the neck like a **necklace**; a bold reddish-orange rump, visible in flight

Where you'll find them:
open country, vacant fields, along railroad tracks, driveways, gravel pits, and wetland edges

Calls and songs:
gives a very loud and distinctive "kill-deer" **call**

On the move:
fakes a broken wing to draw intruders away from the nest, and then takes flight once the nest is safe

What they eat:
loves bugs; also eats worms and snails

Nest:
ground; Dad makes just a slight depression in gravel, often very hard to see

Eggs, chicks, and childcare:
3–5 tan eggs with brown marks; Dad and Mom incubate the eggs and lead the **hatchlings** to food

Spends the winter:
in southern states, Mexico, and Central America; non-migrator in parts of the state

REAL QUICK

Size
11"

Nest
GROUND

Feeder
NONE

year-round
summer

SAW IT!

STAN'S COOL STUFF

Scientists group the Killdeer in the family of shorebirds, but you're more likely to spot one along railroad tracks, around farms, and in other dry habitats than you would at the lakeshore. It's the only shorebird with two black neckbands. It migrates in small flocks.

Brown Thrasher

Look for the long, rusty-red tail

What to look for:
rusty-red bird with a long tail and a long, curved bill, a heavily streaked chest and belly, two white wing bars, and bright yellow eyes

Where you'll find them:
thick shrubs, suburban yards, and forest edges

Calls and songs:
sings a lot and has many different songs; repeats each phrase in the songs twice

On the move:
quickly flies or runs on the ground in and out of dense shrubs

What they eat:
insects and fruit

Nest:
constructs a cup nest low in a dense shrub or evergreen tree, or in uncultivated **vegetation** along a fence (fencerow)

Eggs, chicks, and childcare:
4–5 speckled pale blue eggs; Mom and Dad sit on the eggs and feed their young

Spends the winter:
migrates to southern states; non-migrator in parts of the state

REAL QUICK

Size
11"

Nest
CUP

Feeder
NONE

year-round
summer
migration

SAW IT!

STAN'S COOL STUFF

The male thrasher is a fantastic songster and sings more songs than any other bird in North America—more than 1,100 tunes in all! It's a noisy feeder due to its habit of turning over leaves, small rocks, and branches to find food.

Northern Flicker

Look for the black mustache

What to look for:
brown-and-black bird with a black mustache and black **necklace**, a speckled chest, and a red spot on the nape of neck; female lacks a black mustache

Where you'll find them:
forests, small woods, backyards, parks

Calls and songs:
gives a loud "wacka-wacka" **call**

On the move:
flies in a deep, exaggerated up-and-down pattern, flashing yellow under its wings and tail

What they eat:
insects (especially ants and beetles); known to eat the eggs of other birds; visits **suet** feeders

Nest:
cavity in a tree or in a nest box that is stuffed with sawdust; often reuses an old nest several times

Eggs, chicks, and childcare:
5–8 white eggs; Mom and Dad incubate the eggs and feed the baby woodpeckers

Spends the winter:
in New York; moves around to find food

REAL QUICK

Size
12"

Nest
CAVITY

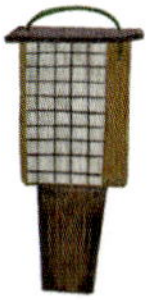

Feeder
SUET

year-round

SAW IT!

STAN'S COOL STUFF

The Northern Flicker is the only woodpecker to regularly feed on the ground. The male often picks the nest site, usually a natural cavity in a tree. Both parents will pitch in to dig as needed, taking as many as 12 days to finish excavating a hole.

Mourning Dove

Look for the shimmering colors on the neck

What to look for:
brown-to-gray bird with shiny, **iridescent** pink and greenish-blue on the neck, a gray patch on the head, and black spots on the wings and tail

Where you'll find them:
around your seed and ground feeders, open fields

Calls and songs:
known for its soft, sad (mournful) cooing

On the move:
wind rushes through its wing feathers during takeoff and flight, creating a whistling sound

What they eat:
seeds; visits ground and seed feeders

Nest:
flimsy platform in a tree, made with twigs; often falls apart in a storm or during high winds

Eggs, chicks, and childcare:
2 white eggs; parents incubate the eggs and feed a regurgitated liquid to their young for the first few days of life

Spends the winter:
in New York and southern states; moves around to find food

REAL QUICK

Size
12"

Nest
PLATFORM

Feeder
GROUND

year-round

SAW IT!

STAN'S COOL STUFF

This dove is a ground feeder that bobs its head as it walks. It's one of the few birds that drinks without lifting its head, like the Rock Pigeon (pg. 171). Parents **regurgitate** a liquid, called **crop-milk**, to feed to their young (**squab**) during their first few days of life.

Pied-billed Grebe

Look for the black ring around the bill

What to look for:
small brown waterbird with a black ring on a thick ivory bill and a puffy white patch under the tail; winter **plumage** has a brown bill

Where you'll find them:
wetlands, ponds, and lakes

Calls and songs:
a rhythmic series of very loud whooping **calls**

On the move:
often dives while swimming, entirely submerging itself to catch food; it can reappear on the surface far from where it went under

What they eat:
crayfish, aquatic insects, and fish

Nest:
ground nest floating in water

Eggs, chicks, and childcare:
5–7 bluish-white eggs; Mom and Dad sit on the eggs and tend to the baby grebes

Spends the winter:
on Long Island, in southern states, Mexico, and Central America

REAL QUICK

Size
12–14"

Nest
GROUND

Feeder
NONE

SAW IT!

STAN'S COOL STUFF

The grebe is well suited to aquatic life. It has short wings, lobed toes, and legs set close to the rear of its body, making it awkward on land. When it's been disturbed, it compresses its feathers to get the air out and then sinks underwater like a submarine.

Blue-winged Teal

Look for the white crescent on the face

MALE

FEMALE

What to look for:

male is brown with black speckles, has a gray head with a white crescent on the face, a white patch on the tail, and a blue wing patch (**speculum**); female is duller and lacks a white facial crescent and white patch on the tail

Where you'll find them:

wetlands and lakes

Calls and songs:

male makes a high-pitched squeak; female quacks

On the move:

direct flight to and from water; **flocks** fly fast in tight formation; female performs a **display** to draw predators away from the nest and young

What they eat:

aquatic plants, seeds, and aquatic bugs

Nest:

cozy ground nest some distance from the water

Eggs, chicks, and childcare:

8–11 creamy-white eggs; Mom does the **incubation** and feeds the ducklings

Spends the winter:

in southern states, Mexico, and Central America

REAL QUICK

Size
15–16"

Nest
GROUND

Feeder
NONE

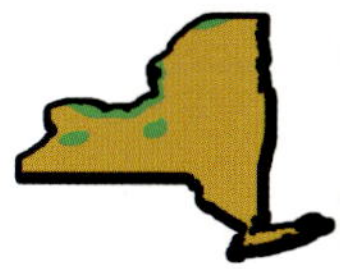

summer
migration

SAW IT!

STAN'S COOL STUFF

This is one of the smallest ducks in North America. It **migrates** farther than most other ducks, nesting as far north as Alaska. The name "Blue-winged" refers to its blue wing patch, which is easiest to see when the bird is in flight.

Red-shouldered Hawk

Look for the reddish shoulders

What to look for:
cinnamon-red head, shoulders, chest and belly; brown wings and back with white spots, and a long tail with black-and-white bands; reddish wing linings, seen in flight

Where you'll find them:
wooded backyards, forest edges, woodlands

Calls and songs:
extremely vocal; gives distinctive, loud screams

On the move:
alternates flapping with gliding

What they eat:
reptiles, amphibians, large insects, and small birds

Nest:
large platform made of sticks and lined with sprigs of evergreen or other soft materials; usually in a fork of a large tree

Eggs, chicks, and childcare:
2–4 white eggs with dark marks; Mom and Dad sit on the eggs and provide for the youngsters

Spends the winter:
in parts of New York; moves around to find food

REAL QUICK

Size
15–19"

Nest
PLATFORM

Feeder
NONE

year-round
summer

SAW IT!

STAN'S COOL STUFF

A common hawk in New York. It likes to hunt along forest edges and will search for snakes, frogs, bugs, and other **prey** as it perches. It stays in the same territory for many years. The parents start to build a nest in March. The young leave the nest (**fledge**) by July.

Hooded Merganser

Look for the ragged "hair" on the head

What to look for:
sleek brown-and-rust bird with ragged "hair" on the back of a red head

Where you'll find them:
wooded areas near ponds, shallow lakes, rivers

Calls and songs:
female gives a hoarse quack; male gives a deep, rolling **call**

On the move:
quick, low flight with fast wingbeats across the water; usually in small groups

What they eat:
small fish, aquatic insects, and **crustaceans** (especially crayfish)

Nest:
cavity; adds a lining of feathers in a natural cavity, an old woodpecker hole or a nest box

Eggs, chicks, and childcare:
10–12 white eggs; Mom incubates the eggs and feeds the babies

Spends the winter:
on Long Island, southern states

REAL QUICK

Size
16–19"

Nest
CAVITY

Feeder
NONE

year-round
summer

SAW IT!

STAN'S COOL STUFF

The merganser is not a duck—it's a small diving bird. The female will lay her eggs in the nests of other mergansers or Wood Ducks (pg. 127), making a total of 20–25 eggs in some nests! Rarely, she will share a nest cavity and sit side by side with a Wood Duck.

Wood Duck

Look for the bright white eye-ring

FEMALE

MALE
pg. 185

What to look for:
brown with a bold white eye-ring; crest on head and blue wing patch (**speculum**) are harder to see

Where you'll find them:
quiet, shallow ponds, and deep in the woods, high up on tree branches

Calls and songs:
female calls "oo-eek, oo-eek" when startled and at takeoff; male calls a zipper-like "zeeeet"

On the move:
blasts off from the water with loud calls and noisy wings; flies quickly through forest; enters cavity nest from full flight; small, tight group flights

What they eat:
aquatic insects, plants, and seeds

Nest:
cavity; adds a lining of soft, downy feathers in an old woodpecker hole or a nest box

Eggs, chicks, and childcare:
10–15 creamy-white eggs; only Mom incubates the eggs and shows the kids how to feed

Spends the winter:
in southern states

REAL QUICK

Size
17–20"

Nest
CAVITY

Feeder
NONE

summer

SAW IT!

STAN'S COOL STUFF

This is a small dabbling duck. The female will lay some eggs in a neighbor's nest (**egg dumping**), sometimes resulting in 20 or more eggs in a nest! The young stay in the nest for a day, then jump from as high as 60 feet to the ground or water and follow their mom.

Common Goldeneye

Look for the yellow-tipped dark bill

FEMALE

MALE
pg. 51

What to look for:
brown body duck with a large brown puffy head, dark bill with yellow tip, yellow eye

Where you'll find them:
on small and large bodies of water, fresh and saltwater

Calls and songs:
female gives incomplete quack; wings whistle while in flight

On the move:
very fast straight flight, constantly flapping

What they eat:
aquatic plants, insects, fish, and mollusks

Nest:
cavity; adds a lining of feathers in an old woodpecker cavity or natural cavity in a tree

Eggs, chicks, and childcare:
8–10 light green eggs; Mom incubates the eggs and leads the babies to food

Spends the winter:
some stay in New York; southern states and Mexico; moves around to find open water

REAL QUICK

Size
18–20"

Nest
CAVITY

Feeder
NONE

summer
migration
winter

SAW IT!

STAN'S COOL STUFF

A small duck of both fresh and saltwater. One of the last duck species to migrate each fall. A diving duck to feed on aquatic insects. Fast flying bird reaching around 45 mph. Found around the planet in northern hemisphere (circumpolar).

Northern Shoveler

Look for the large, shovel-like bill

What to look for:
brown duck with black speckles, green wing mark (**speculum**), and a super-large, spoon-shaped bill

Where you'll find them:
shallow wetlands, ponds, and small lakes

Calls and songs:
female gives a classic quack; male gives a crazy-sounding combination of popping and quacking, calling "puk-puk, puk-puk, puk-puk"

On the move:
swims in tight circles, stirring up insects to eat; small flocks of 5–10 birds swim with bills pointing toward the water; flocks fly in tight formation

What they eat:
enjoys aquatic insects; likes plants, too

Nest:
ground; Mom gathers plant material and forms it into a circle a short distance from the water

Eggs, chicks, and childcare:
9–12 olive eggs; Mom sits on the eggs and leads her little shovelers to food

Spends the winter:
in southern states, Mexico, and Central America; a few winter in parts of New York

REAL QUICK

Size
19–21"

Nest
GROUND

Feeder
NONE

migration
winter

SAW IT!

STAN'S COOL STUFF

The Northern Shoveler is a medium-sized duck. It is the only shoveler species found in North America. The name "Shoveler" refers to its peculiar, shovel-like bill. It feeds by using its bill to sift tiny aquatic insects and plants floating on the water's surface.

Mallard

Look for the orange-and-black bill

FEMALE

What to look for:
overall brown duck with an orange-and-black bill, a white tail, and a blue-and-white wing mark (**speculum**), seen best in flight

Where you'll find them:
lakes and ponds, rivers and streams, and maybe even your backyard

Calls and songs:
the sound a duck makes is based on the female Mallard's classic quack; the male doesn't quack

On the move:
sometimes in huge flocks with hundreds of ducks; mostly in small flocks of 6–10, especially in spring

What they eat:
seeds, aquatic plants, and insects; visits ground feeders offering corn

Nest:
ground; Mom builds it from plants nearby

Eggs, chicks, and childcare:
7–10 greenish-to-whitish eggs; Mom incubates the eggs and leads the young to food

Spends the winter:
in New York; moves around to find food

REAL QUICK

Size
19–21"

Nest
GROUND

Feeder
GROUND

year-round

SAW IT!

STAN'S COOL STUFF

This is a dabbling duck, tipping forward in shallow water to feed on aquatic plants on the bottom. Only the female quacks. It will return to its birthplace each year. The Mottled Duck (see inset) looks very much like the female Mallard except for the bill color.

Red-tailed Hawk

Look for the rusty-red tail

What to look for:
plumage varies but it's often brown with a white chest, brown belly band, and a rusty-red tail

Where you'll find them:
just about anywhere; open country, where it flies over open fields and roadsides; cities, where it perches on freeway light posts, fences, and trees

Calls and songs:
gives a high-pitched scream that trails off

On the move:
hunts in flight, flying in circles as it searches for **prey**; migrates during the day

What they eat:
mice and other animals, birds, snakes, large bugs

Nest:
large platform made of sticks, lined with materials such as evergreen needles; often in a large tree

Eggs, chicks, and childcare:
2–3 white eggs, sometimes speckled; parents sit on the eggs and provide for the youngsters

Spends the winter:
in New York

REAL QUICK

Size
19–23"

Nest
PLATFORM

Feeder
NONE

year-round

SAW IT!

STAN'S COOL STUFF

This **raptor** is a large hawk with a wide variety of colors from bird to bird, ranging from chocolate to nearly all white. The red color of the tail develops in the second year of life and usually is best seen from above. It returns to the same nest site each year.

Barred Owl

Look for the dark eyes

What to look for:
brown-to-gray owl with dark brown eyes, dark horizontal bars on upper chest, vertical streaks on the lower chest and belly; yellow bill and feet

Where you'll find them:
dense woodlands

Calls and songs:
gives calls of 6–8 hoots, sounding something like "who-who-who-cooks-for-you"

On the move:
a smooth and silent flight, gliding on flat, outstretched wings; often hunts during the day, perching and watching for mice and other **prey**

What they eat:
small mammals, birds, fish, reptiles, amphibians

Nest:
natural cavity in a tree or uses a nest box with a large entrance hole; doesn't add nesting material

Eggs, chicks, and childcare:
2–3 white eggs; Mom sits on the eggs; Mom and Dad attend to the babies

Spends the winter:
doesn't **migrate**; stays in New York year-round

REAL QUICK

Size
20–24"

Nest
CAVITY

Feeder
NONE

year-round

SAW IT!

STAN'S COOL STUFF

The Barred Owl is the only common dark-eyed owl in New York. It's a chunky bird with a large head. It fishes by hovering over water and then reaches down to grab one. After the young **fledge**, they stay with their parents for up to four months.

Great Horned Owl

Look for the feather tufts on the head

What to look for:
"eared" owl with large yellow eyes, a V-shaped white throat, and horizontal barring on the chest

Where you'll find them:
just about any **habitat** throughout New York

Calls and songs:
calls a familiar "hoo-hoo-hoo-hoooo"

On the move:
flies silently on big wings that stretch out to 4 feet; takes a few quick flaps, and then glides

What they eat:
small to medium mammals, birds (especially ducks), snakes, and insects

Nest:
no nest; takes over the nest of another bird or uses a broken tree stump or other semi-cavity

Eggs, chicks, and childcare:
2–3 white eggs, laid in January and February; Mom incubates; Dad and Mom feed the **hatchlings**

Spends the winter:
doesn't **migrate** and usually hangs around the same area year after year

REAL QUICK

Size
21–25"

Nest
NONE

Feeder
NONE

year-round

SAW IT!

STAN'S COOL STUFF

The "**horns**" of the Great Horned are feather tufts, not ears. Its eyelids close from the top down, like ours. It has fabulous hearing and can hear a mouse moving under a deep pile of leaves. It's one of the few animals that will kill a skunk or a porcupine.

Ring-necked Pheasant

Look for the long, pointed tail

What to look for:
golden brown with a multicolored head, a white ring around the neck and a super-long tail; female is smaller and less flamboyant, with a shorter tail

Where you'll find them:
open country with shrubs, overgrown strips of land along fences (fencerows), farm fields

Calls and songs:
male gives a cackling **call** to attract females, and then rapidly flutters his wings

On the move:
usually walks or runs; explosive flight with fast wingbeats followed by gliding low to the ground

What they eat:
insects (such as grasshoppers), seeds (like corn kernels), and fruit; visits ground feeders

Nest:
ground; Mom builds the nest

Eggs, chicks, and childcare:
8–10 olive-brown eggs; Mom sits on the eggs and leads her **brood** to food

Spends the winter:
doesn't **migrate**; seeks shelter in harsh weather

REAL QUICK

Size
21–36"

Nest
GROUND

Feeder
GROUND

year-round

SAW IT!

STAN'S COOL STUFF

"Ring-necked" refers to the white ring around the male's neck. Introduced from China to North America over 100 years ago, it's now common in the U.S. Like many other game birds, its numbers vary, making it common in some years and rare in others.

Wild Turkey

Look for the bare blue-and-red head

MALE

FEMALE

What to look for:
funny-looking brown-and-bronze bird, bare blue-and-red head, long thin beard, large fanning tail; female is thinner, duller, and often lacks a beard

Where you'll find them:
just about any **habitat**, from suburban yards to prairies and forests

Calls and songs:
a fast, descending "gobble-gobble-gobble-gobble" that's often heard before the bird is seen

On the move:
a strong flier that can approach 60 miles per hour; also able to fly straight up, and then away

What they eat:
insects, seeds, and fruit

Nest:
ground; Mom scrapes out a shallow depression and pads it with soft leaves

Eggs, chicks, and childcare:
10–12 whitish eggs with dull brown marks; Mom sits on the eggs and leads the babies to food

Spends the winter:
moves around New York to find **cover** and food

REAL QUICK

Size
36–48"

Nest
GROUND

Feeder
NONE

year-round

SAW IT!

STAN'S COOL STUFF

The turkey is the largest game bird in New York. The male's head and neck change color when displaying for females. It sees three times better than people, and it can hear sounds from a mile away. A male will lead a group of up to 20 females.

Ruby-crowned Kinglet

Look for the ruby crown

What to look for:
small, teardrop-shaped green-to-gray bird; two white wing bars and a white eye ring; hidden ruby crown

Where you'll find them:
thick shrubs low to the ground

Calls and songs:
a distinctive song that starts out soft and ends loud and on a higher note

On the move:
quickly flits from one low branch to another; rarely stays still

What they eat:
insects and berries; will come to peanut or suet feeders if no insects available

Nest:
pendulous; Mom builds it

Eggs, chicks, and childcare:
4–5 white eggs with brown marks; Mom incubates the eggs and Dad helps her feed the kids

Spends the winter:
on Long Island, in southern states, Mexico, and Central America

REAL QUICK

Size
4"

Nest
PENDULOUS

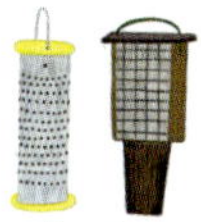

Feeder
TUBE OR SUET

summer
migration
winter

SAW IT!

STAN'S COOL STUFF

One of the smallest birds in New York. The red crown (ruby) is often hidden until the bird is excited. Early spring migrant, often traveling in flocks with other birds.

Black-capped Chickadee

Look for the black cap

What to look for:
mostly gray bird with a black cap and throat patch, tan sides and belly, and a white chest

Where you'll find them:
nearly all habitats—just look around for this bird

Calls and songs:
calls "chika-dee-dee-dee-dee"; also gives a high-pitched, two-toned "fee-bee" **call** during spring; can have different calls in different regions

On the move:
flies short distances with short, fluttery wings

What they eat:
seeds, bugs, and fruit; visits seed and **suet** feeders

Nest:
excavates a cavity or uses a nest box; gathers mostly green moss for the nest and fur to line it

Eggs, chicks, and childcare:
5–7 white eggs with fine brown marks; Mom and Dad sit on the eggs and feed their **brood**

Spends the winter:
doesn't **migrate**; moves around to find food and shelter instead; must feed every day in winter and forages for food even during the worst snowstorms

REAL QUICK

Size
5"

Nest
CAVITY

Feeder
TUBE OR SUET

year-round

SAW IT!

STAN'S COOL STUFF

You can attract this bird with a seed feeder or nest box. Usually, it's the first to find a new feeder. It is easily tamed and hand-fed. Most of the diet comes from bird feeders, so it can be a common urban bird. It's often seen with nuthatches, woodpeckers, and other birds.

White-breasted Nuthatch

Look for the white chest

MALE

FEMALE

What to look for:
gray back with a white face, chest and belly, a black cap and nape of neck, and a large white patch on the rump; female has a gray cap and **nape**

Where you'll find them:
woodlands, parks, backyards, forest edges

Calls and songs:
a characteristic spring **call**, "whi-whi-whi-whi," given during February and March

On the move:
climbs down tree trunks headfirst, looking for hidden bugs; quick, short flights from tree to tree

What they eat:
bugs, bug eggs, seeds; visits seed and **suet** feeders

Nest:
cavity; Mom and Dad build a nest in an empty woodpecker hole or a natural cavity

Eggs, chicks, and childcare:
5–7 white eggs with brown marks; Mom sits on the eggs; Mom and Dad feed the little ones

Spends the winter:
doesn't **migrate** in New York; moves around to find food

REAL QUICK

Size
5–6"

Nest
CAVITY

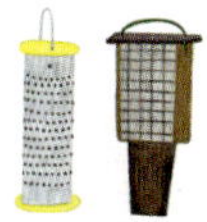

Feeder
TUBE OR SUET

year-round

SAW IT!

STAN'S COOL STUFF

This is one of almost 30 nuthatch species worldwide. It has an extra-long hind toe claw, called a nail, on each foot, giving it the ability to cling to trees and climb down headfirst. It's often in flocks with chickadees. Pairs stay together all year and defend their territory.

Yellow-rumped Warbler

Look for the bright yellow patches

What to look for:
gray with black streaks on the chest and yellow patches on head, flanks, and rump; female is duller gray; first-winter juvenile is similar to the female

Where you'll find them:
can be seen in any **habitat** during migration; seems to prefer **deciduous** forests and forest edges

Calls and songs:
sings a wonderful song in spring; calls a single robust "chip," heard mostly during migration

On the move:
quickly moves among trees and from the ground to trees; flits around upper branches of tall trees

What they eat:
insects and berries; visits **suet** feeders in spring

Nest:
cup; Mom builds the nest on her own in forests

Eggs, chicks, and childcare:
4–5 white eggs with brown marks; Mom sits on the eggs; Mom and Dad feed the young

Spends the winter:
to southern states; a few stay on Long Island

REAL QUICK

Size
5–6"

Nest
CUP

Feeder
SUET

summer
winter

SAW IT!

STAN'S COOL STUFF

This bird is also called the Myrtle Warbler. One of the most common warblers in the state. The male molts his gray feathers in fall, changing to a dull color like the female for the winter. He keeps his yellow patches all year.

Dark-eyed Junco

Look for the pink bill

What to look for:
plump bird with a gray-to-charcoal chest, head and back, a white belly, and a tiny pink bill

Where you'll find them:
on the ground in small flocks with other juncos and sparrows

Calls and songs:
a beautiful, loud musical **trill** lasting 2–3 seconds

On the move:
outermost tail feathers are white and appear as a white V during flight

What they eat:
seeds (scoffs down many weed seeds) and insects; visits ground and seed feeders

Nest:
cup on the ground in a wide variety of habitats; female chooses a well-hidden nest site

Eggs, chicks, and childcare:
3–5 white eggs with reddish-brown marks; Mom incubates the eggs; Dad and Mom feed the babies

Spends the winter:
stays in parts of New York; moves around in winter

REAL QUICK

Size
5½"

Nest
CUP

Feeder
GROUND

year-round
winter

SAW IT!

STAN'S COOL STUFF

The junco is one of our most common winter birds. Females **migrate** farther south than males. This round, dark-eyed bird uses both feet at the same time to "**double-scratch**" the ground, exposing seeds and insects to eat.

Tufted Titmouse

Look for the pointed crest

What to look for:
gray bird with a white chest and belly, a rusty-brown wash on the flanks and a pointed crest

Where you'll find them:
woodlands, backyards, and parks

Calls and songs:
quickly repeats a loud "peter-peter-peter" **call**

On the move:
usually seen only one or two at a time, never in big flocks, moving through thick forests and along forest edges

What they eat:
insects, seeds (especially black oil sunflower seeds), and fruit; visits seed and **suet** feeders

Nest:
cavity; boldly pulls hair from sleeping dogs, cats, or squirrels and uses it to line an old woodpecker hole or a nest box

Eggs, chicks, and childcare:
5–7 white eggs with brown marks; Mom sits on the eggs, and Mom and Dad feed the young

Spends the winter:
doesn't **migrate**; stays in New York year-round

REAL QUICK

Size
6"

Nest
CAVITY

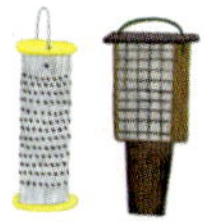

Feeder
TUBE OR SUET

year-round

SAW IT!

STAN'S COOL STUFF

The titmouse is a common feeder bird. You can attract it by filling a feeder with black oil sunflower seeds or peanuts or by putting up a nest box. The male feeds the female during courtship and nesting. Its name means "small bird."

Eastern Phoebe

Look for the tail pumping while perching

What to look for:
plain gray with darker wings, a light olive-green belly, and a thin dark bill

Where you'll find them:
end of dead branches in forests, yards, and farms

Calls and songs:
repeats a characteristic "fee-bee" **call** over and over from the top of dead branches

On the move:
waits on a branch for an insect to pass by, flies out to snatch it, and returns to the branch—called **hawking**; bobs its tail up and down when perched

What they eat:
insects

Nest:
cup made of mud, grass, and green moss, lined with hair (and sometimes feathers); under a bridge or eave of a house, or in another sheltered area

Eggs, chicks, and childcare:
4–5 white eggs; Mom sits on the eggs to incubate; Dad and Mom feed their **hatchlings**

Spends the winter:
in southern states and Mexico

REAL QUICK

Size
7"

Nest
CUP

Feeder
NONE

summer

SAW IT!

STAN'S COOL STUFF

The Eastern Phoebe is named for its very distinct call, "fee-bee." Because it is a plain bird that doesn't have distinctive markings, it's easier to identify by its hawking and tail-pumping behaviors, and by the call that sounds like its name.

Sanderling

Look for the gray back

pg. 101

What to look for:
a light-colored sandpiper with a gray head and back, white belly, and black legs and bill; white stripe on the wings, seen only in flight

Where you'll find them:
on sandy beaches

Calls and songs:
gives a fast, high-pitched squeak just before taking flight and just after takeoff

On the move:
when waves retreat at the beach, groups run out to feed; often hops away from people on one leg; performs a distraction **display** when threatened

What they eat:
insects, crabs, worms, and small **mollusks**

Nest:
ground nest; Dad builds it

Eggs, chicks, and childcare:
3–4 greenish-olive eggs with brown marks; the parents do the **incubation** and feed the kids

Spends the winter:
on East and Gulf Coasts, in the Caribbean, Mexico, Central and South America

REAL QUICK

Size
8"

Nest
GROUND

Feeder
NONE

SAW IT!

STAN'S COOL STUFF

The Sanderling is one of the most common shorebirds in New York. It has winter **plumage** from August to April and breeding plumage from April to August. To rest, it stands on one leg and tucks the other leg into its belly feathers. It nests in the Arctic.

Eastern Screech-Owl

Look for the feathered "ear" tufts

What to look for:
small "eared" owl with two different speckled colorations: gray-and-white and red-and-white

Where you'll find them:
any type of forest, as long as it has some natural cavities suitable for roosting and nesting

Calls and songs:
a trembling, descending **trill**, like a sound effect in a scary movie; seldom gives a screeching **call**

On the move:
flies silently on short, rapidly flapping wings; active from dusk to dawn

What they eat:
large bugs, small mammals, birds, and snakes

Nest:
cavity; uses an old woodpecker hole or a wooden nest box and doesn't add nesting material

Eggs, chicks, and childcare:
4–5 white eggs; Mom sits on the eggs to incubate; Dad feeds Mom and the newly hatched babies

Spends the winter:
moves around the state in winter

REAL QUICK

Size
8–10"

Nest
CAVITY

Feeder
NONE

year-round

SAW IT!

STAN'S COOL STUFF

This little owl has excellent hearing and eyesight. On winter days, it will sun itself at a nest box entrance hole. The male and female may roost with each other at night and may have a long-term **pair bond**. The gray **morph** is more common than the red variety.

Gray Catbird

Look for the black crown

What to look for:
gray bird with a black crown and long, thin black bill; often lifts its tail, exposing a chestnut patch

Where you'll find them:
thick shrubs, forest edges, backyards, parks

Calls and songs:
gives a rasping **call** that sounds like a house cat meowing; often mimics other birds

On the move:
quickly zips back into shrubs if approached

What they eat:
insects and occasional fruit; visits suet feeders

Nest:
cup of small twigs in thick shrubs; often adds a scrap of plastic to the twigs

Eggs, chicks, and childcare:
4–6 blue-green eggs; Mom incubates the eggs; Mom and Dad take turns feeding the chicks, but because the parents look the same, it's hard to tell who's doing the feeding!

Spends the winter:
to southern states; a few stay on Long Island

REAL QUICK

Size
9"

Nest
CUP

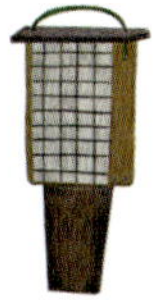

Feeder
SUET

year-round
summer

SAW IT!

STAN'S COOL STUFF

This handsome, secretive bird is more often heard than seen. The Chippewa Indian name for it means "the bird that cries with grief." Once you've heard the call, you won't forget it. If a cowbird lays an egg in a catbird nest, the catbird will break it and eject it.

American Robin

Look for the rusty-red breast

What to look for:
black head and a rich, rusty-red breast; female is duller with a gray head and lighter breast

Where you'll find them:
loves to hop on lawns in search of worms

Calls and songs:
chips and chirps; sings all night in spring; studies report that city robins sing louder than country robins so they can be heard over traffic and noise

On the move:
found all over the U.S. in an amazing range of habitats from sea level to mountaintops

What they eat:
insects, fruit, and berries, as well as earthworms

Nest:
cup; weaves plant materials and uses mud to plaster the nest to a sheltered location

Eggs, chicks, and childcare:
4–7 pale blue eggs; Mom sits on the eggs; Mom and Dad feed the baby robins

Spends the winter:
in New York; moves around in winter to find food; some move to southern states

REAL QUICK

Size
9–11"

Nest
CUP

Feeder
NONE

year-round
summer

SAW IT!

STAN'S COOL STUFF

When a robin walks across your lawn and turns its head to the side, it isn't listening for worms—it is looking for them. Because its eyes are on the sides of its head, a robin must focus its sight out of one eye to see the dirt moving caused by a moving worm.

Northern Mockingbird

Look for the white wing patches

What to look for:
silver-gray head and back, a light gray chest and belly, white wing patches, a mostly black tail with white outer tail feathers

Where you'll find them:
on top of a shrub, where it sits for long periods; parks and yards

Calls and songs:
imitates or mocks other birds (vocal mimicry); young males often sing at night

On the move:
very lively, spreading its wings, flashing its white wing patches, and wagging its tail; wing patches flash during flight or **display**

What they eat:
insects and fruit

Nest:
cup; Mom and Dad work together to build it

Eggs, chicks, and childcare:
3–5 speckled blue-green eggs; Mom sits on the eggs to incubate; Mom and Dad feed their young

Spends the winter:
in much of New York; moves around to find food

REAL QUICK

Size
10"

Nest
CUP

Feeder
NONE

year-round
summer

SAW IT!

STAN'S COOL STUFF

Mockingbirds perform a fantastic mating dance. Pairs hold up their heads and tails and run toward each other. They flash their wing patches and then retreat to nearby **cover**. Usually, they're not afraid of people, so you may be able to get a close look.

Eurasian Collared-Dove

Look for the black collar on the back of neck

What to look for:
pale gray bird with a slightly darker back, wings and tail, and a black collar on the **nape** of neck; tail is long and squared off at the end

Where you'll find them:
wherever it can scratch for seeds

Calls and songs:
gives a series of coos, often during a **display**

On the move:
usually found in small to large flocks; fast flaps, then glides with wings in a V

What they eat:
seeds and fruit; visits ground and seed feeders

Nest:
platform on a building, balcony, barn or shed, or under a bridge; Mom and Dad construct it

Eggs, chicks, and childcare:
3–5 creamy-white eggs; Mom and Dad sit on the eggs and **regurgitate** a liquid food, called **crop-milk**, to feed to the young (**squab**) during their first few days of life

Spends the winter:
doesn't **migrate**; stays in New York year-round

REAL QUICK

Size
12½"

Nest
PLATFORM

Feeder
GROUND

year-round

SAW IT!

STAN'S COOL STUFF

This non-native dove is originally from Asia. People brought it into the Bahamas, and then it flew into Florida in the 1980s. Since then, it has spread across the country. Scientists believe that this bird will continue to expand its range across more of North America.

Rock Pigeon

Look for the gleaming, iridescent patches

What to look for:
color pattern varies; usually shades of gray with gleaming, **iridescent** patches of green mixed with blue; often has a light rump patch

Where you'll find them:
nearly anyplace where it can scratch for seeds

Calls and songs:
a series of coos, usually given during a **display**

On the move:
typically in small to large flocks; flaps rapidly, then glides with wings in a V shape

What they eat:
seeds and fruit; visits ground and seed feeders

Nest:
platform on a building, balcony, barn or shed, or under a bridge

Eggs, chicks, and childcare:
1–2 white eggs; Mom and Dad sit on the eggs and **regurgitate** a liquid, called **crop-milk**, to feed to the young (**squab**) for their first few days of life

Spends the winter:
doesn't **migrate**; stays in New York year-round

REAL QUICK

Size
13"

Nest
PLATFORM

Feeder
GROUND

year-round

SAW IT!

STAN'S COOL STUFF

The pigeon was introduced to North America by the early settlers from Europe. Years of breeding in captivity have made it one of the few birds that has a variety of colors. It's also one of the few birds that can drink without tilting its head back.

Cooper's Hawk

Look for the banded, rounded tail

What to look for:
gray back, rusty chest, short wings, dark red eyes, and a long tail with black bands and a rounded tip

Where you'll find them:
variety of habitats, from woodlands to backyards and parks

Calls and songs:
calls a loud, clear "cack-cack-cack-cack"

On the move:
flies with a few quick flaps followed by long glides

What they eat:
small birds (hunts birds at feeders) and mammals

Nest:
platform made with sticks and leaves, secured high in a tree; Dad and Mom build it together

Eggs, chicks, and childcare:
2–4 greenish eggs with brown marks; parents take turns sitting on the eggs and feeding the young

Spends the winter:
stays year-round in New York; moves around to find food

REAL QUICK

Size
14–20"

Nest
PLATFORM

Feeder
NONE

year-round

SAW IT!

STAN'S COOL STUFF

The stubby wings help this medium-sized hawk move around trees while it chases smaller birds. It will ambush **prey**, flying into brush and running after the birds that flee. **Fledglings** have gray eyes that turn yellow at 1 year and dark red after 3–5 years.

Peregrine Falcon

Look for the yellow eye-ring

What to look for:
dark gray back, tan-to-white chest, dark **hood** head marking, wide black mustache, yellow eye-ring, base of bill and legs, horizontal bars on belly, legs, and undertail; female is similar to male but noticeably larger

Where you'll find them:
canyons, open country and cities with tall buildings

Calls and songs:
alarm **call** is a series of high-pitched screams

On the move:
fast direct flight with constant wingbeats

What they eat:
birds, such as Rock Pigeons in cities and shorebirds and **waterfowl** in rural areas

Nest:
ground (scrape) on a cliff edge, tall building, bridge or smokestack

Eggs, chicks, and childcare:
3–4 white eggs with brown marks; parents take turns sitting on the eggs and feeding the young

Spends the winter:
stays in New York; moves around to find food in winter

REAL QUICK

Size
16–20"

Nest
GROUND

Feeder
NONE

year-round
summer
migration

SAW IT!

STAN'S COOL STUFF

Peregrine Falcons were once nearly eliminated due to indiscriminate shooting and trapping. Now, they are doing well in most parts of the country. This is the fastest bird on the planet. It can fly 70 miles per hour in level flight and can dive up to 200 miles per hour.

Canada Goose

Look for the white cheek strap

What to look for:
large gray goose with a black neck and head, and a white chin and cheek strap

Where you'll find them:
wetlands, ponds, lakes, rivers, and just about any **habitat** with some water

Calls and songs:
belts out its classic "honk-honk-honk," especially during flight

On the move:
flies in a **flock** in a large V shape when traveling long distances

What they eat:
aquatic plants, insects, and seeds

Nest:
ground nest of **vegetation** formed into a mound, usually very near or on water

Eggs, chicks, and childcare:
5–10 white eggs; Mom incubates the eggs; young follow the parents around and learn what to eat

Spends the winter:
in New York; moves around to find open water in winter

REAL QUICK

Size
25–43"

Nest
GROUND

Feeder
NONE

year-round

SAW IT!

STAN'S COOL STUFF

Males guard the flock, bobbing their heads and hissing whenever people approach. Adults **molt** their flight feathers while raising the young, making families temporarily flightless. Canada Geese start to breed in their third year of life. Adults mate for many years.

Great Blue Heron

Look for the long yellow bill

What to look for:
tall gray heron with black eyebrows that end in plumes off the back of the head, neck feathers that drop down like a **necklace**, and a long yellow bill

Where you'll find them:
open water, from small ponds to large lakes

Calls and songs:
when startled, it barks repeatedly like a dog and keeps at it while flying away

On the move:
holds its neck in an S shape in flight and slightly cups its wings, trailing its legs straight out behind

What they eat:
small fish, frogs, insects, snakes, and baby birds

Nest:
platform in a tree near or over open water, in a **colony** of up to 100 birds

Eggs, chicks, and childcare:
3–5 blue-green eggs; parents incubate the eggs and feed the **brood**

Spends the winter:
in southern states, Mexico, Central America, and South America; some stay in New York

REAL QUICK

Size
42–48"

Nest
PLATFORM

Feeder
NONE

year-round
summer

SAW IT!

STAN'S COOL STUFF

This is the tallest heron in New York and very common. It stalks fish in shallow water and strikes at mice, squirrels, and anything else it can capture on land. Red-winged Blackbirds (pg. 25) often attack it to prevent it from taking their babies out of their nests.

Ruby-throated Hummingbird

Look for the gleaming ruby-red throat

What to look for:
tiny **iridescent** green bird with a black throat patch that shimmers bright ruby-red in direct sunlight; female lacks the throat patch

Where you'll find them:
many habitats, from yards and parks to forests

Calls and songs:
will chatter or buzz to communicate; doesn't sing or hum a melody—it's the incredibly fast flapping wings that create the humming sound

On the move:
the only bird that can fly backward; also hovers in midair and flies straight up and straight down!

What they eat:
nectar and insects; visits nectar feeders

Nest:
stretchy cup of plant materials and spiderwebs; glues bits of **lichen** to the outside for camouflage

Eggs, chicks, and childcare:
2 white eggs; Mom does all egg and chick care

Spends the winter:
in southern states, Mexico, and Central America

REAL QUICK

Size
3–3½"

Nest
CUP

Feeder
NECTAR

summer

SAW IT!

STAN'S COOL STUFF

This is the tiniest bird in the state, with about the same weight as a U.S. penny. It flaps 50–60 times per second or more when flying at top speed. It breathes 250 times per minute, and its heart beats 1,260 times per minute! It feeds at colorful tube-shaped flowers.

Green Heron

Look for the rusty-red chest

What to look for:
short, stocky heron with a blue-green back, a dark green crest, and a rusty-red chest and neck; short orange legs turn yellow after the breeding season

Where you'll find them:
ponds, wetlands, small lakes, and rivers

Calls and songs:
often gives an explosive, rasping "skyew" **call** when startled

On the move:
waits on the shore or wades stealthily, hunting for food; makes short, quick flights across the water

What they eat:
small fish, aquatic insects, amphibians

Nest:
platform of sticks in a **coniferous** or **deciduous** tree, often a short distance from water and can be very high in the tree

Eggs, chicks, and childcare:
2–4 light green eggs; parents share the childcare

Spends the winter:
in southern states, Mexico, Central America, and South America

REAL QUICK

Size
16–22"

Nest
PLATFORM

Feeder
NONE

summer
migration

SAW IT!

STAN'S COOL STUFF

The Green Heron holds its head very close to its body. When it's excited, it raises its crest. To attract fish to catch, it will place an object, such as an insect, on the surface of the water. Baby herons make a loud ticking sound, like the ticktock of a clock.

Wood Duck

Look for the boldly patterned head

What to look for:
boldly patterned head and crest with bold white outlines; rusty chest and white belly

Where you'll find them:
quiet, shallow ponds and deep in the woods, high up on tree branches

Calls and songs:
male calls a zipper-like "zeeeet"; female calls "oo-eek, oo-eek" loudly when startled and at takeoff

On the move:
blasts off from the water with loud calls and noisy wings; flies quickly through forest; enters cavity nest from full flight; small, tight group flights

What they eat:
aquatic insects, plants, and seeds

Nest:
cavity; adds a lining of soft, downy feathers in an old woodpecker hole or a nest box

Eggs, chicks, and childcare:
10–15 creamy-white eggs; only Mom incubates the eggs and shows the kids how to feed

Spends the winter:
in southern states

REAL QUICK

Size
17–20"

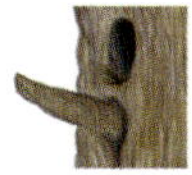

Nest
CAVITY

Feeder
NONE

summer

SAW IT!

STAN'S COOL STUFF

This is a small dabbling duck. The female will lay some eggs in a neighbor's nest (**egg dumping**), sometimes resulting in 20 or more eggs in a nest! The young stay in the nest for 24 hours, and then jump down to follow their mother. They never return to the nest.

Northern Shoveler

Look for the large, shovel-like bill

MALE

FEMALE
pg. 131

What to look for:
shiny, **iridescent** green head with rusty sides, a white chest, and a super-large, spoon-shaped bill

Where you'll find them:
shallow wetlands, ponds, and small lakes

Calls and songs:
male gives a crazy-sounding combination of popping and quacking, calling "puk-puk, puk-puk, puk-puk"; female gives a classic quack **call**

On the move:
swims in tight circles, stirring up insects to eat; small flocks of 5–10 birds swim with bills pointing toward the water; flocks fly in tight formation

What they eat:
enjoys aquatic insects; likes plants, too

Nest:
ground; Mom gathers plant material and forms it into a circle a short distance from the water

Eggs, chicks, and childcare:
9–12 olive eggs; Mom sits on the eggs and leads her little shovelers to food

Spends the winter:
in southern states, Mexico, and Central America; a few winter in parts of New York

REAL QUICK

Size
19–21"

Nest
GROUND

Feeder
NONE

migration
winter

SAW IT!

STAN'S COOL STUFF

The Northern Shoveler is a medium-sized duck. It is the only shoveler species found in North America. The name "Shoveler" refers to its peculiar, shovel-like bill. It feeds by using its bill to sift tiny aquatic insects and plants floating on the water's surface.

Mallard

Look for the green head

What to look for:

green head with a white **necklace**, rusty-brown chest, gray sides, yellow bill, orange legs, and feet

Where you'll find them:

lakes and ponds, rivers and streams, and maybe even your backyard

Calls and songs:

the male doesn't quack; when you think of how a duck sounds, it's based on the female Mallard's classic loud quack

On the move:

sometimes in huge flocks with hundreds of ducks; mostly in small flocks of 6–10, especially in spring

What they eat:

seeds, aquatic plants, and insects; visits ground feeders offering corn

Nest:

ground; Mom builds it from plants nearby

Eggs, chicks, and childcare:

7–10 greenish-to-whitish eggs; Mom incubates the eggs and leads the young to food

Spends the winter:

in New York; some migrate to southern states

REAL QUICK

Size
19–21"

Nest
GROUND

Feeder
GROUND

year-round

SAW IT!

STAN'S COOL STUFF

This is a dabbling duck, tipping forward in shallow water to eat plants on the bottom. The male has black feathers in the center of its tail that curl upward. The common name "Mallard" means "male" and refers to the males, which don't help raise their young.

Baltimore Oriole

Look for the black head

pg. 217

What to look for:
flaming orange bird with a black head and back, and black wings with white wing bars

Where you'll find them:
parks, yards, and forests; in treetops, where it feeds on caterpillars

Calls and songs:
a fantastic songster, singing loudly; often heard before it is seen

On the move:
often returns to the same area year after year

What they eat:
insects, fruit and **nectar**; comes to nectar, orange-half, and grape jelly feeders

Nest:
pendulous; an interesting nest that looks like a sock hanging from an outer branch of a tall tree

Eggs, chicks, and childcare:
4–5 bluish eggs with brown marks; Mom sits on the eggs; Mom and Dad do the childcare

Spends the winter:
in southern states, Mexico, Central America, and South America

REAL QUICK

Size
7–8"

Nest
PENDULOUS

Feeder
NECTAR

SAW IT!

STAN'S COOL STUFF

Orioles come to feeders that offer sugar water (nectar), orange halves, or grape jelly. Parents bring their young to feeders. Young males turn orange and black at 1½ years. Members of the blackbird family; closely related to grackles and Red-winged Blackbirds.

House Finch

Look for the reddish face and the brown cap

What to look for:
red-to-orange face, throat, chest, and rump, and a brown cap

Where you'll find them:
forests, city, and suburban areas, around homes, parks and farms

Calls and songs:
male sings a loud, cheerful warbling song

On the move:
moves around in small family units; never travels long distances

What they eat:
seeds, fruit, and leaf buds; comes to seed feeders and feeders with a glop of grape jelly

Nest:
cup, but occasionally in a cavity; likes to nest in a hanging flower basket or on a front door wreath

Eggs, chicks, and childcare:
4–5 pale blue eggs, lightly marked; Mom sits on the eggs and Dad feeds her while she incubates; Mom and Dad feed the **brood**

Spends the winter:
in New York; moves around to find food

REAL QUICK

Size
5"

Nest
CUP

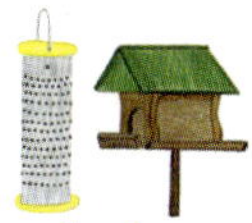

Feeder
TUBE OR HOPPER

year-round

SAW IT!

STAN'S COOL STUFF

The House Finch is very social and is found across the country. It can be the most common bird at feeders. Unfortunately, it suffers from an eye disease that causes the eyes to crust over. It's rare to see a yellow male; yellow **plumage** may be a result of a poor diet.

Scarlet Tanager

Look for the black wings

What to look for:
tropical-looking scarlet-red bird with coal-black wings and tail, and an ivory bill

Where you'll find them:
mature **deciduous** woodlands

Calls and songs:
both the male and female sing songs similar to the American Robin (pg. 165), but they sprinkle in an unusual "chick-burr" **call** in their tunes; the female song is like the male's, only softer

On the move:
flies high up around treetops, hunting for insects; arrives late in spring and leaves early in fall

What they eat:
insects and fruit

Nest:
cup in a territory of 4–8 acres with large patches of forest

Eggs, chicks, and childcare:
4–5 speckled blue-green eggs; Mom incubates the eggs; Mom and Dad feed the babies

Spends the winter:
in Central and South America

REAL QUICK

Size
7"

Nest
CUP

Feeder
NONE

summer

SAW IT!

STAN'S COOL STUFF

There are about 240 tanager species, and nearly all have bright colors and live in the tropics. "Tanager" is from a South American Indian word meaning "any small, brightly colored bird." After the male sheds (molts) his red **plumage** in fall, he looks like the female.

Northern Cardinal

Look for the black mask

What to look for:
all-red bird with a black mask, and a large red crest and bill

Where you'll find them:
wide variety of habitats including backyards and parks; usually likes thick **vegetation**

Calls and songs:
calls "whata-cheer-cheer-cheer" in spring; both male and female sing and give chip notes all year

On the move:
short flights from **cover** to cover, often landing on the ground

What they eat:
loves sunflower seeds and enjoys insects, fruit, peanuts, and **suet**; visits seed feeders

Nest:
cup of twigs and bark strips, often low in a tree

Eggs, chicks, and childcare:
3–4 speckled bluish-white eggs; Mom and Dad share the incubating and feeding duties

Spends the winter:
doesn't **migrate**; gathers with other cardinals and moves around to find good sources of food

REAL QUICK

Size
8–9"

Nest
CUP

Feeder
TUBE OR HOPPER

year-round

SAW IT!

STAN'S COOL STUFF

Cardinals are sunbathers! Sometimes, they stretch out in the sun, spreading their wings and fanning their tails. They're the first to arrive at feeders in the morning and the last to leave before dark. They are territorial and fight their own reflections in windows.

Laughing Gull

Look for the black head "hood"

What to look for:
black head "**hood**" with a white neck, chest, and belly, gray wings with black wing tips, and orange bill; winter **plumage** head is gray-and-white with a black bill

Where you'll find them:
along the coasts; freshwater and saltwater sites

Calls and songs:
a loud series of calls that sound like laughter; male tosses his head back and calls to attract a mate

On the move:
almost always in groups, moving from one water **habitat** to another

What they eat:
fish, insects on land and in the water

Nest:
ground nest, lined with grass, sticks, and rocks; nests in a marsh in a large **colony**

Eggs, chicks, and childcare:
2–4 olive eggs with brown marks; parents sit on the eggs and **regurgitate** food to feed the young

Spends the winter:
in New York; many migrate to southern states

REAL QUICK

Size
16–17"

Nest
GROUND

Feeder
NONE

year-round
winter

SAW IT!

STAN'S COOL STUFF

It takes this gull a few years to get adult plumage. During the first year, the young start out mostly brown-and-gray. They look like adults during the second year, but they don't have an all-black head "hood." Juveniles get the breeding plumage in the third year.

Ring-billed Gull

Look for the black ring on the bill

What to look for:
white gull with gray wings and a yellow bill with a black ring near the tip; winter **plumage** has speckles on the head and neck

Where you'll find them:
shores of large lakes and rivers; often at garbage dumps and parking lots

Calls and songs:
calls out a wide variety of loud, rising squawks and squeals—classic gull sounds

On the move:
strong flight with constant wing flaps

What they eat:
insects and fish; it also picks through garbage, scavenging for other food

Nest:
ground; defends a small area around it

Eggs, chicks, and childcare:
2–4 off-white eggs with brown marks; Mom and Dad take turns incubating the eggs and feeding their young

Spends the winter:
in New York; many migrate to southern states

REAL QUICK

Size
18–20"

Nest
GROUND

Feeder
NONE

year-round
migration

SAW IT!

STAN'S COOL STUFF

This is one of the most common gulls in the country. Hundreds of these birds often **flock** together. The ring on the bill appears after the first winter. In the fall of the first 3 years, the birds have a different plumage. In the third year, they grow adult plumage.

Great Egret

Look for the long, thin white neck

REAL QUICK

Size
36–40"

Nest
PLATFORM

Feeder
NONE

summer
migration

What to look for:
tall and thin white egret with a long neck, long legs, and a long, pointed yellow bill

Where you'll find them:
shallow wetlands, ponds, and lakes

Calls and songs:
gives a loud, dry croak if disturbed or when it squabbles for a nest site at the **colony**

On the move:
holds its neck in an S shape during flight; slowly stalks in shallow water, looking for fish to spear with its sharp bill

What they eat:
small fish, aquatic insects, frogs, and crayfish

Nest:
platform, in a colony of up to 100 birds

Eggs, chicks, and childcare:
2–3 light blue eggs; Mom and Dad sit on the eggs and give food to the **hatchlings**

Spends the winter:
in southern states, Mexico, and Central America

SAW IT!

STAN'S COOL STUFF

From the 1800s to the early 1900s, the Great Egret was hunted to near extinction for its beautiful long plumes, which were used to decorate women's hats. The plumes grow near the tail during the breeding season. Today, the egret is a protected bird.

Mute Swan

Look for the black knob at base of bill

What to look for:
large all white swan with distinctive orange bill, large black knob at base of bill

Where you'll find them:
any water in the state

Calls and songs:
a hoarse and muffled call with a sputtery end

On the move:
strong and direct flight

What they eat:
aquatic insects and plants

Nest:
ground; Mom and Dad build it

Eggs, chicks, and childcare:
4–8 light gray eggs; Mom and Dad take turns sitting on the eggs

Spends the winter:
in New York; moves around in winter to find open water and food

REAL QUICK

Size
60"

Nest
GROUND

Feeder
NONE

year-round

SAW IT!

STAN'S COOL STUFF

Even though it's called a mute swan, it still makes sounds. They defend their territory aggressively, resulting in many skirmishes with other swans. A non-native bird that was brought in as a swan for parks and estates. The male is called a "cob" and female a "pen."

American Goldfinch

Look for the black forehead

What to look for:
bright canary-yellow bird with a black forehead, wings, and tail; female is olive-yellow and lacks a black forehead; winter male resembles the female

Where you'll find them:
open fields, scrubby areas, woodlands, backyards

Calls and songs:
male sings a pleasant high-pitched song; gives **twitter** calls during flight

On the move:
appears roller-coaster-like in flight

What they eat:
loves seeds and insects; comes to seed (especially thistle) feeders

Nest:
cup; builds its nest in late summer and lines the cup with the soft, silky down from wild thistle

Eggs, chicks, and childcare:
4–6 pale blue eggs; Mom incubates the eggs and Dad pitches in to help her feed the babies

Spends the winter:
in New York and southern states; flocks of up to 20 birds move around in winter to find food

REAL QUICK

Size
5"

Nest
CUP

Feeder
TUBE OR HOPPER

year-round

SAW IT!

STAN'S COOL STUFF

The American Goldfinch is often called Wild Canary due to its canary-colored **plumage**. This feeder bird is almost always in small flocks, visiting thistle tube feeders that offer Nyjer seed. A late-nesting bird with most nesting in July through mid-September.

Common Yellowthroat

Look for the black mask

MALE

FEMALE

What to look for:
olive-brown with a bold yellow throat and chest, and a thick black mask; female lacks the mask

Where you'll find them:
open fields with tall grass, marshes with a thick tangle of **vegetation**

Calls and songs:
a cheerful "witchity-witchity-witchity-witchity" song, sung from deep within tall grass; male also sings from prominent perches and when he hunts

On the move:
male bounces in and out of tall grass in a curious courtship **display** while singing a mating song

What they eat:
bugs, all sizes

Nest:
cup, built low to the ground

Eggs, chicks, and childcare:
3–5 white eggs with brown marks; Mom sits on the eggs; Mom and Dad feed bugs to the babies

Spends the winter:
flies all the way to southern states, Mexico, and Central America

REAL QUICK

Size
5"

Nest
CUP

Feeder
NONE

summer

SAW IT!

STAN'S COOL STUFF

The Common Yellowthroat is a common warbler throughout New York. The male has a distinctive black mask outlined in white. It breeds across the state during summer. The young stay dependent on their parents longer than most other warblers.

Yellow Warbler

Look for the orange streaks on the chest

MALE

FEMALE

What to look for:
yellow bird with thin orange streaks on the chest and belly; female lacks orange streaks

Where you'll find them:
gardens and shrubby areas near water, backyards

Calls and songs:
male sings a string of sweet notes, sounding like "sweet, sweet, sweet, I'm-so-sweet!"

On the move:
zooms around shrubs and shorter trees; begins to migrate south in August, migrating at night in mixed flocks of warblers; males return to claim territories 1–2 weeks before females arrive

What they eat:
insects

Nest:
cup; Mom builds it

Eggs, chicks, and childcare:
4–5 white eggs with brown marks; Mom sits on the eggs; Mom and Dad give food to the kids

Spends the winter:
in southern states, Mexico, Central America, and South America

REAL QUICK

Size
5"

Nest
CUP

Feeder
NONE

summer

SAW IT!

STAN'S COOL STUFF

The Yellow Warbler is common and widespread in New York. It eats small caterpillars and many other bugs on tree leaves. The male is easier to see higher up in trees than the duller female. He sings loudly, zips off to grab a bug, and then starts singing again.

Palm Warbler

Look for the yellow eyebrows and chestnut ca

What to look for:
yellow eyebrows, throat, and belly, and yellow under tail; brown crown; a dark line across dark eyes

Where you'll find them:
woodlands

Calls and songs:
a muffled and uneven trill, which is usually heard during migration

On the move:
hops rather than walks; look for it to wag or bob its tail while catching insects from the leaves and flowers of trees

What they eat:
insects and fruit

Nest:
cup; Mom builds it

Eggs, chicks, and childcare:
4–5 white eggs with brown marks; Mom incubates the eggs; Mom and Dad feed the young

Spends the winter:
in southeastern coastal states, the Caribbean, and Central America

REAL QUICK

Size
5½"

Nest
CUP

Feeder
GROUND

migration

SAW IT!

STAN'S COOL STUFF

When the Palm Warbler recognizes cowbird eggs in its nest, it builds a new nest over the top of the cowbird eggs, keeping them from being incubated. It is often seen near the ground feeding on insects.

Scarlet Tanager

Look for the olive wings

What to look for:
greenish-yellow bird with olive wings and tail, and an ivory bill

Where you'll find them:
mature **deciduous** woodlands

Calls and songs:
both the female and male sing songs similar to the American Robin (pg. 165), but they sprinkle in an unusual "chick-burr" **call** in their tunes; the female song is like the male's, only softer

On the move:
flies high up around treetops, hunting for insects; arrives late in spring and leaves early in fall

What they eat:
insects and fruit

Nest:
cup in a territory of 4–8 acres with large patches of forest

Eggs, chicks, and childcare:
4–5 speckled blue-green eggs; Mom incubates the eggs; Mom and Dad feed the babies

Spends the winter:
in Central and South America

REAL QUICK

Size
7"

Nest
CUP

Feeder
NONE

summer

SAW IT!

STAN'S COOL STUFF

There are about 240 tanager species and nearly all have bright colors and live in the tropics. "Tanager" is from a South American Indian word meaning "any small, brightly colored bird." The male looks like the female after he sheds (molts) his red **plumage** in fall.

Baltimore Oriole

Look for the gray-brown wings

What to look for:
pale yellow bird with orange tones and gray-brown wings with white wing bars

Where you'll find them:
parks, yards, and forests; in treetops, where it feeds on caterpillars

Calls and songs:
a fantastic songster, singing loudly; often heard before it is seen

On the move:
often returns to the same area year after year

What they eat:
insects, fruit, and **nectar**; comes to nectar, orange-half, and grape jelly feeders

Nest:
pendulous; an interesting nest that looks like a sock hanging from an outer branch of a tall tree

Eggs, chicks, and childcare:
4–5 bluish eggs with brown marks; Mom sits on the eggs; Mom and Dad do the childcare

Spends the winter:
in Mexico, Central America, and South America

REAL QUICK

Size
7–8"

Nest
PENDULOUS

Feeder
NECTAR

summer

SAW IT!

STAN'S COOL STUFF

Orioles come to feeders that offer sugar water (nectar), orange halves, or grape jelly. Parents bring their young to feeders. Young males look like females for their first 1½ years. They're some of the last birds to arrive in spring and some of the first to leave in fall.

Eastern Meadowlark

Look for the V-shaped black necklace

What to look for:
robin-shaped bird with a short tail, a yellow chest and belly, and a V-shaped black **necklace**; white outer tail feathers, usually seen when flying away

Where you'll find them:
meadows, open grassy country, roadsides

Calls and songs:
sings a wonderful flute-like, clear whistling song

On the move:
if you move toward it when it's perching on a fence post, it will quickly dive into tall grass

What they eat:
insects and seeds

Nest:
cup on the ground in dense **cover**; Mom builds the nest by herself

Eggs, chicks, and childcare:
3–5 white eggs with brown marks; Mom sits on the eggs, but both parents feed the **hatchlings**; pairs have 2 broods per year

Spends the winter:
in southern states; some don't **migrate**; moves around in winter to find food

REAL QUICK

Size
9"

Nest
CUP

Feeder
NONE

year-round

SAW IT!

STAN'S COOL STUFF

The Eastern Meadowlark is a bird that likes to live in meadows. Meadowlarks are ground-dwelling songbirds that sing while they fly. They are members of the blackbird family, which makes them relatives of orioles and grackles.

BIRD FOOD FUN FOR THE FAMILY

If you and your family like to do fun projects together, making your own bird food and bird-feeding items might be just the right ones to try. Chances are good that you already have most of the ingredients at home to make delicious and nutritious treats for your wild bird friends.

You'll be doing these projects in the kitchen, so show your mom or dad the following sections. They're written specifically with the whole family in mind. For example, you may need to check with a parent or guardian for help with such tasks as grocery shopping, stovetop cooking, or food preparation, like cutting up fresh fruit.

Starter Snacks and Fruit Treats

You can start by offering some food that's already in your kitchen. Peanut butter attracts a lot of birds! Simply use a spatula to smear some on the bark of a nearby tree where you can watch it from a window. Or use a piece of firewood: prop it up or hang it with a rope and slather it with peanut butter—then watch the birds go wild.

To offer treats like raisins, dates, and currants, place them outside in a nonbreakable small bowl with a few holes drilled in the bottom for drainage. Waxwings, robins, and many other birds love small dried fruit, and some will be flying in shortly to get some.

Putting out fresh fruit, such as apples and oranges, is another great way to attract bright and colorful birds to your yard. Cut these into small, manageable pieces, and offer the snacks on the tray of a feeder.

Another cool way to serve an orange is to cut one in half and place the halves sunny-side up on a feeder or branch. This arrangement allows birds to easily feast on the sweet fruit contained inside the rind. Sometimes, it's best to impale the orange half on a nail to stop it from rolling away.

Plain and unsalted nuts, especially peanuts, pecans, and walnuts, make wonderful treats for birds. Simply add these to a feeder tray with seeds or place them in a tube feeder for nuts.

Easy Bird Food Recipes

Preparing bird food of any kind shows that you care about the birds in your backyard. Now, are you ready to try making some recipes? Below are just a few suggestions. You can find much more online.

Sweet Homemade Nectar

Nectar is a superb food for many birds. Studies of nectar from flowers have shown that the average flower nectar is 25 percent sucrose. Sucrose is a simple sugar, so to make the correct strength of homemade nectar (sugar water), mix a ratio of 1 part sugar to 4 parts water. You'll discover that hummingbirds, orioles, and woodpeckers will thoroughly enjoy the sweet drink that you made.

INGREDIENTS
¼ cup granulated white sugar
1 cup warm water

DIRECTIONS: Add the sugar to the water and stir to dissolve. If you prefer, you can boil the water first so the sugar dissolves more quickly. Cool to room temperature before filling your feeder. Store any extra in the refrigerator or freezer.

NOTES: Never substitute brown sugar or honey for white sugar. Also, there is no need to add red food coloring because the birds will be attracted to any amount of red on any part of your **nectar** feeder.

Birds-Go-Wild Spread

INGREDIENTS
½ cup raisins
½ cup granola
½ cup oatmeal
½ cup Cheerios
16-ounce jar smooth peanut butter

DIRECTIONS: Mix dry ingredients in a large mixing bowl. Warm the peanut butter in a microwave or place the jar in warm water to soften. Scoop out the softened peanut butter and mix well with dry ingredients until smooth.

Spread on tree bark or smear a few dollops on the tray of a feeder.

Love-It-Nutty Butter

INGREDIENTS
2 cups shelled peanuts, unsalted
2 cups shelled walnuts, unsalted
¼ cup raisins
3–5 tablespoons coconut oil or other vegetable oil

DIRECTIONS: Toss dry ingredients into a food processor. Start blending. Add oil until the mixture reaches a smooth, thick consistency. Store in refrigerator.

Smear on a wooden board with grooves or spread on tree bark.

Make Your Own Suet

You and your family can make outstanding **suet** recipes at home, too. Suet is animal fat, often from cows, and there are several convenient ways to get it for your recipes.

A quick way is to purchase plain suet cakes. In store-bought suet, the fat has already been melted down (**rendered**). A cheaper way might be to buy fat trimmings in bulk from your local butcher or large amounts of **lard** at your grocery store. A clever way to get rendered fat from your own kitchen is for an adult to pour fat drippings from cooked bacon, pork, and beef into an empty, clean can. When the fat has cooled and solidified, cover and refrigerate to save for future use.

Easy-Peasy Suet

INGREDIENTS
1 cup solidified fat of your choice
1 cup chunky peanut butter
3 cups ground cornmeal
1 cup white flour
1 cup black oil sunflower seeds or peanuts

DIRECTIONS: In a large pot, melt the fat over low heat. Do not heat quickly, or the fat might burn. Strain the fat through a **cheesecloth** to remove any chunks, and then pour the liquid back into the pot.

Add the peanut butter to the fat. Stir over low heat until the mixture melts and consistency is smooth. Remove from heat. Add the cornmeal and flour and mix until stiff. Add the sunflower seeds or peanuts, and mix thoroughly.

Pour into a mold or container. With a spatula, spread out the mixture and smooth the top. Cool completely, then cut into squares. Store in freezer.

Simply Super Suet

INGREDIENTS
2 cups **suet** or **lard**
1 cup peanut butter
2 cups yellow cornmeal
2 cups cracked corn
1 cup black oil sunflower seeds

DIRECTIONS: In a large pot, melt the suet or lard over low heat. Add the peanut butter, stirring until melted and well mixed. Add remaining ingredients, and mix.

Pour into baking pans or forms and allow to cool. Cut into chunks or shapes. Store in freezer.

Yummy Bird-Feeding Projects

Bird-feeding projects are super activities for families, and they can be a big hit at special occasions, such as birthday parties. These very attractive ornaments and feeders also make unique gifts for the holidays and family celebrations.

Birdseed Ornaments

INGREDIENTS
cookie cutters in any shape
nonstick cooking spray
½ cup water
3 tablespoons white corn syrup
2½ teaspoons unflavored gelatin
¾ cup white flour
4 cups black oil sunflower seeds
10- to 12-inch pieces of string or **jute** twine

DIRECTIONS: Place the cookie cutters on wax paper and spray with nonstick cooking spray. Set aside.

In a saucepan, bring the water and corn syrup to a boil. Reduce heat and stir in gelatin until completely mixed. Do not overcook.

Transfer the hot liquid to a bowl. Add the flour and mix until smooth. Add the sunflower seeds and mix well. Mixture will now be thick.

Use a spatula to fill each cookie cutter. Be sure to press the seeds into all parts of the shapes. Roll any extra mixture into balls. Poke one hole through each shape and each ball with a pencil or similar object.

When cooled, pop out the ornaments from the cookie cutters. Thread a length of string or twine through each hole and tie the ends to form a loop. Loop each of your ornaments over nearby branches, and watch the birds come to feast!

Pine Cone Birdseed Feeders

Try your hand at making this fabulous little bird feeder from an ordinary pine cone. It's fun and easy, and everyone in your family can make their own.

INGREDIENTS (per person)
1 pine cone
10- to 12-inch piece of string or **jute** twine
peanut butter
birdseed

DIRECTIONS: Tie a piece of string or twine to a pine cone. Roll the cone in peanut butter, filling the spaces between the "petals" (bracts) and coating the entire surface. Then, roll the cone in birdseed until the seeds completely cover the peanut butter.

Hang the feeder outside where you can see the birds feeding on it, and enjoy the show!

MORE ACTIVITIES FOR THE BIRD-MINDED

Nothing brings family and friends closer together than a shared interest. Birding and backyard bird feeding are enjoyable, year-round activities that many find appealing. Here are some things to do that are not only fun for everyone but also supportive of the birds.

Help Birds Build Their Nests

A thoughtful way for the entire family to work together with birds during spring is to put out a variety of soft and flexible natural items to help birds build their nests.

First, gather some everyday materials around your home that birds will use. Here are some excellent items to offer:

- Yarn, cut into 6-inch-long pieces
- Fabric from an old, clean T-shirt, cut into 6-inch-long strips
- **Cotton batting** (used for handicrafts)
- Fuzzy pet hair from a brush

Next, place your materials into an unused, clean **suet** cage. Be sure to let the ends of the yarn and fabric strips hang out, and don't pack the material in tightly. The birds need to be able to take out the items easily.

Hang the cage by a short chain from a tree in early spring, when the birds are starting to construct their nests. And then, wait...

Soon, birds will be flying back and forth to the materials and choosing their favorites. It's a delight to see birds making use of your nesting contributions. Not only have you assisted the bird parents, but you've also helped them provide a comfy home for their families. Good job!

Make a Bird-watching List

Making a watch list on **poster board** of the birds that have visited your yard is a handicraft project that the whole family will enjoy. You can decorate the poster any way you like, but it's awesome to show pictures of the birds you've spotted and write notes about the sightings.

Each time you see a new species in your yard, mark it on the poster with the date and time of day. Attach it to the refrigerator, or put it in another prominent place where it's easy for everyone in the family to see and add their updates.

Your watch list is also a valuable way to track the arrival of the first hummingbirds and orioles in your area each spring. If you create a new watch list each year, it could reveal trends in the arrival dates. This information would be of interest not only to your family and friends but also to your teachers and local birding organizations.

Save the Birds with Hawk Cutouts

Another fun and important project is to make hawk cutouts to attach to your windows. These items will help prevent birds from flying into sheets of glass at your home.

In-flight window strikes are one of the major killers of our wild bird friends. Window reflections of the sky, trees and other natural features in your yard create the illusion for birds that the flight path is clear. When birds see forms of predator birds in the reflections, they will turn away and take another route.

Various web pages show outlines (**silhouettes**) of hawks that you can print and cut out. Check the possibilities, and then pick your favorites.

Tape the cutouts to any large picture windows, as well as other windows and doors with clear glass. This preventive action will greatly reduce the risk of birds crashing headfirst into glass. Then, give yourself a high five for helping to save them.

Build Your Very Own Birdhouse

A first-rate project for kids and adults to do together is to construct a birdhouse. Building plans are available online for different kinds of birdhouses for different kinds of birds. Give them a once-over, and pick one that you like best for the birds you want nesting nearby.

The instructions online will help you select the right kind of wood and show you how to cut it to the right sizes. Most importantly, the plans will provide the correct size of the entrance hole for the bird, along with how-to instructions for making it. Most birdhouse projects require hand and power tools, so be sure to work with an adult.

You might even want to make multiple birdhouses with your extended family or your neighbors. With everyone doing different tasks, your team can turn out a bluebird box, a wren box, a robin platform, and more!

Create a Bird-Friendly Yard

There is no better way to support the birds in your area than to plant bird-friendly flowers, bushes, and trees. There are many varieties of these plants, making it easy to choose some that will be ideal for your yard.

Planting perennials that bloom large and showy flowers each year is an outstanding way to feed hummingbirds. Many shrubs produce attractive **nectar**-filled flowers and then, later in the summer, edible fruit, which the birds love. Numerous tree species offer berries and nuts—foods the birds depend on in late fall.

A yard with grass alone just isn't a friendly **habitat** for birds, so sit down with your family and think about putting in a flower garden or adorning your yard with some shrubs and trees. Soon afterward, you'll be hearing the sweet chirping of birds and a rich repertoire of **birdsong** all around you.

Take a Birding Trip

Everyone loves a good time! For a fun family outing, plan a birding trip to a local park, state park, or national wildlife refuge. In spring, you'll be rewarded with migrating warblers. During summer, all of the nesting birds will be feeding babies. In fall, **waterfowl** will be superactive. Even in winter, there are many amazing birds to see.

Your local nature center is another good place to see birds. Oftentimes, nature centers have bird feeders set up to attract birds. Stop in after school or early on Saturday mornings to see what comes to the feeders.

Practice Good Birding

Finding a stray feather or an empty bird nest is exciting when you and your family are sharing time in nature. Examining these wonders and making a sketch or taking photos are always fun educational opportunities. However, everyone should be aware that collecting, possessing, or owning wild bird feathers, nests, and even bird eggs is not permitted under federal law.

It may seem silly that a lost feather or vacant bird nest needs protecting, but very important laws stop people from buying, selling, and trading these items. In the past, a lively market for feathers, bird nests, and also eggs led to widespread killing of birds, some to near extinction. To prevent this from happening again, strong laws were passed to safeguard all of our bird species.

So enjoy seeing, studying, and learning about birds, but please don't take any feathers, nests, or eggs with you out of their natural environment. Leave them just as you found them, and perhaps someone else will also get the opportunity to benefit from studying them.

COMMUNITY SCIENCE PROJECTS

I can't think of a more exciting way to learn about birds and expand the birding experience than to take part in a community science project. If you are unfamiliar with community science projects, they are sponsored by organizations in which citizens like yourself can contribute in a meaningful way to actual scientific projects right from your own home! Most projects don't take much time and can be fun family activities, with everyone sharing what they learned about birds.

There are simple community science projects that might have you just count the birds that come to your feeders. Others are more complex and involve more time, effort, and perhaps a little traveling. Either way, I'm sure you can find an enjoyable and educational community project that will be a perfect fit for your family. Give it a try!

Here are some popular projects and resources for you to explore:

The very well-known Christmas Bird Count winter census, FeederWatch and more

birds.cornell.edu/home

Hummingbird migration

journeynorth.org/tm/humm/AboutSpring.html

Finding and counting nesting birds

nestwatch.org

General community science projects for counting birds

birdwatchingdaily.com/featured-stories/year-round-citizen-science-projects

American Kestrel nesting and population study

kestrel.peregrinefund.org

LEARNING ABOUT BIRDING ON THE INTERNET

Birding online is another fine way to discover more information about birds—plus, it's a terrific way to spend time during rainy summer days and winter evenings after sunset. So check out the websites below, and be sure to share the fabulous things you've learned about birds with your family and friends.

eBird

https://ebird.org/home

American Birding Association: Young Birders

https://www.aba.org/aba-young-birders/

Cornell Lab of Ornithology

www.birds.cornell.edu/home

Author Stan Tekiela's website

www.naturesmart.com

In addition, online birding groups can be of valuable assistance to you as well. Facebook has many pages dedicated to specific areas of the state and the birds that live there. These sites are an excellent, real-time resource that will help you spot birds in your region. Consider joining a Facebook birding group.

GLOSSARY

birdsong: A series of musical notes that a bird strings together in a pleasing melody. Also called a song. See *warble*.

brood: A family of bird brothers and sisters that hatched at around the same time.

brood parasites: Birds that don't nest, incubate or raise families, such as Brown-headed Cowbirds (pg. 21). See *host*.

call: A nonmusical sound, often a single note, repeated. See *note*.

carrion: A dead and often rotting animal's body, or carcass, that is an important food for many other animals, including birds.

cheesecloth: A loosely woven cotton cloth used primarily to wrap cheese but also used to strain particles from liquids.

colony: A group of birds nesting together in the same area. The size of a colony can range from two pairs to hundreds of birds.

coniferous: A tree or shrub that has evergreen, needle-like leaves and that produces cones.

cotton batting: A light, soft cotton material, often used to stuff quilts.

cover: A dense area of trees or shrubs where birds nest or hide.

crop-milk: A liquid that pigeons and doves regurgitate (spit up) to feed their young.

crustaceans: A large, mainly aquatic group of critters, such as crayfish, crabs, and shrimp.

deciduous: A tree or shrub that sheds its leaves every year.

display: An attention-getting behavior of birds to impress and attract a mate, or to draw predators away from the nest. A display may include dramatic movements in flight or on the ground.

double-scratch: A hunting behavior of some ground feeders, such as juncos, involving hopping forward and scratching the ground back with both feet at the same time to uncover seeds.

egg dumping: A nesting behavior of some birds, such as Wood Ducks (pg. 185), in which the female lays some of her eggs in another female's nest in addition to her own nest.

epaulets: Decorative color patches on the shoulders of a bird, as seen in male Red-winged Blackbirds (pg. 25).

excavate: To dig or remove wood or dirt, creating a cavity, hole or tunnel.

fledge: The process of developing flight feathers and leaving the nest.

fledglings: Young birds that have recently left the nest. See *hatchlings*.

flock: A group of the same bird species or a gathering of mixed species of birds. Flocks range from a pair of birds to upwards of 10,000 individuals.

habitat: The natural home or environment of a bird.

hatchlings: Baby birds that have recently emerged from their eggs. See *fledglings*.

hood: The markings on the head of a bird, resembling a hood.

horns: A tuft or collection of feathers, usually on top of a bird's head, resembling horns.

host: A bird species, such as the Red-winged Blackbird (pg. 25), that takes care of the eggs and babies of other bird species. See *brood parasites*.

incubation: The process of sitting on bird eggs in the nest to keep them warm until they hatch.

iridescent: A luminous, or bright, quality of feathers, with colors seeming to change when viewed from different angles.

jute: A string of rough fibers made from plants.

lard: Fat from mammals, such as cows and pigs.

lichen: A unique partnership of plant and fungi growing together and looking and acting as one organism.

lores: The areas on each side of a bird's face between the eye and the base of the bill.

migrate: The regular and predictable pattern of seasonal movement by some birds from one region to another, especially to escape winter.

molt: The process of dropping old, worn-out feathers and replacing them with new feathers, usually only one feather at a time.

mute: The inability to make or produce audible sounds. The Turkey Vulture (pg. 35), for example, is mostly mute.

nape: The back of a bird's neck.

necklace: The markings around the neck of a bird, as seen in the Eastern Meadowlark (pg. 219).

nectar: A sugar and water solution usually consisting of approximately 25 percent sucrose and 75 percent water, usually found in flowers.

nestlings: Young birds that have not yet left the nest. See *hatchlings*.

note: A single sound of a call, such as the "chip" note of a Northern Cardinal (pg. 197), that doesn't change pitch or frequency. See *birdsong*.

pair bond: The relationship between a male and female bird during the mating season.

plumage: The collective set of feathers on a bird at any given time.

poster board: A stiff cardboard used for displaying information.

prey: Any critter that is hunted and killed by another for food.

raptor: A flesh-eating bird of prey that hunts and kills for food. Hawks, eagles, Ospreys, falcons, owls, and vultures are raptors. See *prey*.

regurgitate: The process of bringing swallowed food up again to the mouth to feed young birds.

rendered: Animal fat that has been reduced or melted down by heating in order to make it pure.

silhouettes: Dark shapes or outlines against a lighter background.

speculum: A patch of bright feathers on some birds, such as ducks, found on the wings.

squab: A young pigeon or dove, usually still in the nest.

suet: Animal fat, usually beef, that has been heated and made into cakes to feed birds. See *rendered*.

trachea: A large tubelike organ that allows air to pass between the lung and the mouth of a bird. Also called a windpipe.

tree sap: The watery liquid that moves up and down within the circulatory system of a tree, carrying nutrients throughout.

trill: A fluttering or repeated series of similar-sounding musical notes given by some birds, especially Dark-eyed Juncos (pg. 153). See *note*.

twitter: A high-pitched call of a bird. See *call*.

ultraviolet light: A kind of light that is visible to birds and insects but unseen by people.

vegetation: Any plants, especially those found in a particular habitat.

warble: A series of pleasing musical notes strung together and often changing. See *trill*.

waterfowl: A group of similar birds with a strong connection to water. Ducks, geese, swans, and others, including Blue-winged Teals (pg. 121), are waterfowl.

CHECKLIST/INDEX BY SPECIES

Use the circles to checkmark the birds you've seen.

ABOUT THE AUTHOR

Naturalist, wildlife photographer, and writer Stan Tekiela is the originator of the popular state-specific field guide series that includes *Birds of New York Field Guide*. He has authored more than 195 field guides, nature books, children's books, wildlife audio CDs, and playing cards, presenting many species of birds, mammals, reptiles, amphibians, trees, wildflowers, and cacti in the United States.

With a Bachelor of Science degree in Natural History from the University of Minnesota, and as an active professional naturalist for more than 30 years, Stan studies and photographs wildlife throughout the United States and Canada. He has received various national and regional awards for his books and photographs. Also a well-known columnist and radio personality, his syndicated column appears in more than 25 newspapers, and his wildlife programs are broadcast on a number of Midwest radio stations. Stan can be contacted via his website, naturesmart.com.